Riemann, Topology, and Physics

Second Edition

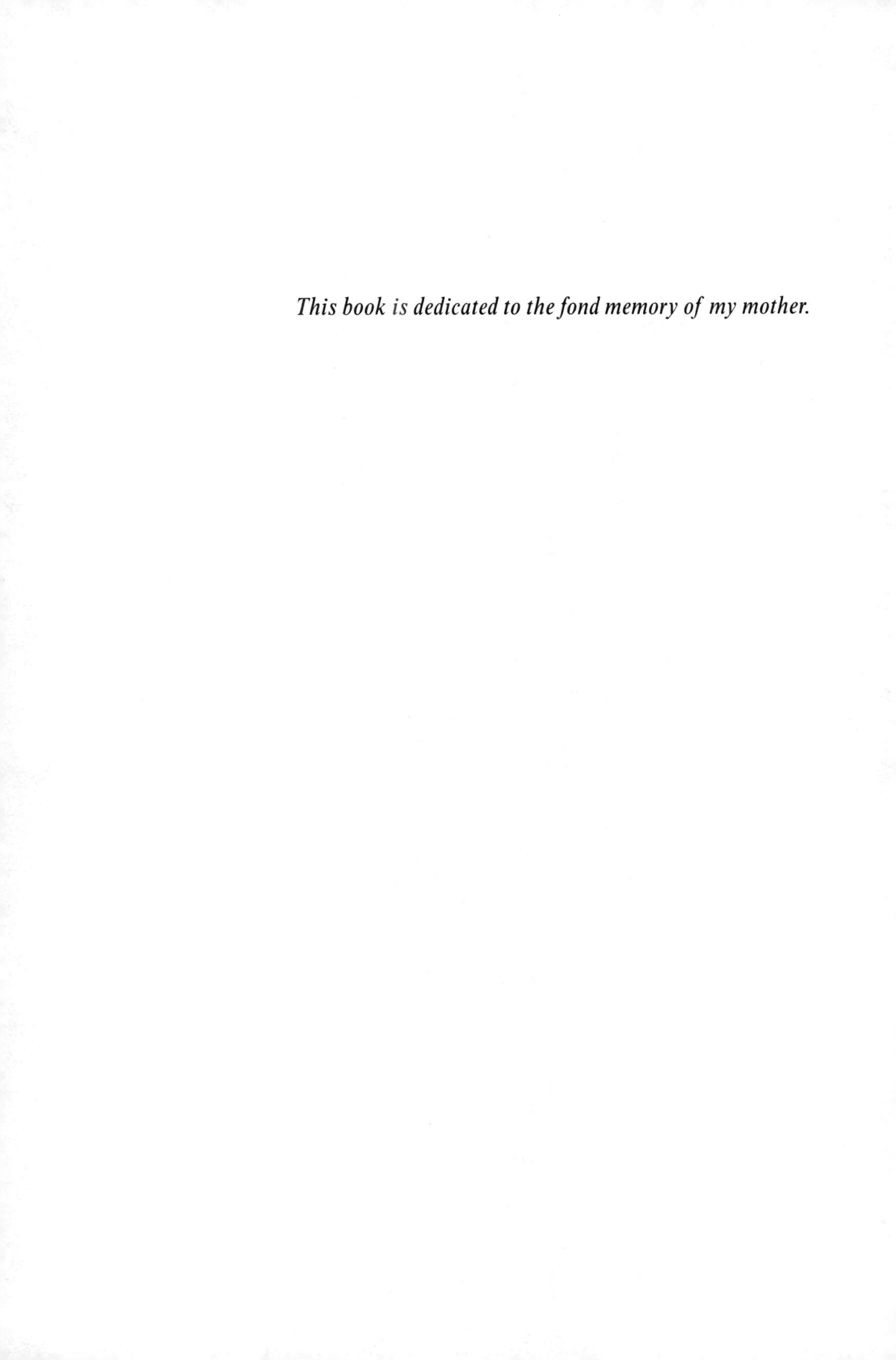

This book is dedicated to the fond memory of my mother.

Michael Monastyrsky

Riemann, Topology, and Physics

Second Edition

With a Foreword by Freeman J. Dyson

Translated by
Roger Cooke
James King
Victoria King

Birkhäuser
Boston • Basel • Berlin

Michael Monastyrsky
Department of Theoretical Physics
Insitute for Theoretical and
Experimental Physics
Moscow 117529, Russia

Roger Cooke (*Translator*)
Department of Mathematics & Statistics
University of Vermont
Burlington, VT 04505

James King (*Translator*)
Victoria King (*Translator*)
7326 55th Avenue N.E.
Seattle, WA 98115

Library of Congress Cataloging-in-Publication Data

Monastyrskiĭ, Mikhail Il'ich.
[Bernkhard Riman. English]
Riemann, topology, and physics / Michael Monastyrsky ; with a foreword by Freeman J. Dyson. — 2nd ed. / translated by Roger Cooke.
p. cm.
Includes bibliographical references and index.
ISBN 0-8176-3789-3 (alk. paper). — ISBN 3-7643-3789-3 (alk. paper)
1. Riemann, Bernhard, 1826--1866. 2. Topology. 3. Mathematical physics. 4. Mathematicians—Germany—Biography. I. Title.
QA29.R425M6613 1998
530.15—dc21 98-7034
CIP

AMS Subject Classifications: 01Axx, 12-02, 12-03, 53-02, 53-03, 55-02, 55-03, 70-02, 70-03

Printed on acid-free paper.

Part I of the 1st edition previously appeared as Bernhard Riemann, © Izdatel'stvo "Znanije," Moscow, 1979.

ISBN 0-8176-3789-3
ISBN 3-7643-3789-3

Reformatted from translator's disk by TEXniques, Inc., Cambridge, MA.
Printed and bound by Quinn-Woodbine, Woodbine, NJ.
Printed in the United States of America.

9 8 7 6 5 4 3 2 1

Contents

Preface to the Second Edition

Ten years have passed since the publication of the first English edition. "Ten years is a long time, even when you are not in jail." If we recall that the article on Riemann appeared in *Priroda* (*Nature*) in 1976, this Russian vaudeville joke is more than appropriate. Only the author's youth can account for the insane enterprise of presenting the scientific achievements and the biography of Riemann in 60 pages and the connection between physics and topology in the same space. Judging from the fact that the book sold out and got favorable reviews, there is a demand for publications of this type. It seems to me that popularizations aimed not at a narrow specialist but at a broader reader are especially needed nowadays. Specialists obtain information in their fields almost instantly thanks to the Internet, but to find out what is happening in contiguous fields, what problems and results are of great interest here, is not at all easy.

Returning to my own book, I note with a certain pride that at least I seem to have evaluated accurately the trends in the development of theoretical physics. It is the combination of physics with topology and algebraic geometry that has led to the brightest achievements of mathematical physics in the last decade. It suffices to note the remarkable results of S. Donaldson on 4-dimensional smooth manifolds obtained in the early 1980's and closely connected with the theory of gauge fields (instantons) and the very recent papers of E. Witten and N. Seiberg, which are also based on physical ideas and are shedding new light on Donaldson's theory.

A second indisputable example is the development of knot theory after the remarkable 1984 paper of V. Jones. Here also the main progress is closely connected with the intertwining of ideas and methods from field theory, statistical physics, and topology. I have tried to include some of the new advances in the second edition, keeping in mind the popular level on which this book is written.

Now a few words about the first part, in which the scientific biography of Riemann is discussed. This part has undergone only minor changes. I have corrected certain historical and mathematical inaccuracies and a large number of misprints, and added a few new facts about the development of Riemann's ideas. Indeed the last decade has confirmed the amazing modernity and fecundity of Riemann's ideas and results.

The most sensational result of recent years has been the proof of Fermat's last theorem. The proof of A. Wiles and R. Taylor uses the full power of modern algebraic geometry, whose foundation played a definitive role for Riemann's ideas and methods.

Of the classical legacy of mathematics only Riemann's hypothesis on the zeros of the ζ-function still remains unproved.

Interest in the personality and works of Riemann has noticeably increased in recent years. It suffices to mention the beautiful new edition of Riemann's works by R. Narasimhan, which contains articles by well-known specialists on the current state of the branches of mathematics in which Riemann worked. A solid biography of Riemann written by D. Laugwitz appeared in German in 1996 and recently in English.

I hope that my little book will be useful to the reader who wishes to look into several branches of mathematics and physics through the prism of the history of the life and works of the man who determined its shape.

M. MONASTYRSKY
December, 1998
Harvard University
Cambridge, MA

Foreword to the First Edition

Soviet citizens can buy Monastyrsky's biography of Riemann for eleven kopeks. This translated volume will cost considerably more, but it is still good value for the money. And we get Monastyrsky's monograph on topological methods in the bargain. It was a good idea of Birkhäuser Boston to publish the two translations in one volume. The economics of publishing in a capitalist country make it impossible for us to produce the small cheap paperback booklets, low in quality of paper and high in quality of scholarship, at which the Soviet publishing industry excels. Monastyrsky's two booklets are outstanding examples of the genre. By putting them together, Birkhäuser has enabled them to fit into the Western book-marketing system.

The two booklets were written separately and each is complete in itself, but they complement each other beautifully. The Riemann biography is short and terse, like Riemann's own writings. It describes in few words and fewer equations the revolutionary ideas which Riemann brought into mathematics and physics a hundred and twenty years ago. The topological methods booklet describes how some of these same ideas, after lying dormant for a century, found new and fruitful applications in the physics of our own time. The two parts of the story together illustrate one of the central themes of science, the mysterious power of mathematical concepts to prepare the ground for physical discoveries which could not have been foreseen or even imagined by the mathematicians who gave the concepts birth. In telling this story, Monastyrsky does not begin, as Dostoevsky began *The Brothers Karamazov*, with a quotation from the Gospel of St. John: "Verily, verily, I say unto you, except a corn of wheat fall into the ground and die, it abideth alone; but if it die, it bringeth forth much fruit." Dostoevsky's epigraph would be as appropriate to the history of mathematical physics as to the history of the human soul. Riemann's grains of wheat, many of them unknown and unpublished during his lifetime, are still bringing forth fruit abundantly today.

The quality which makes Monastyrsky unique among expositors of contemporary physics is his depth of historical focus. He sees modern ideas in a perspective which goes back all the way to Riemann. But this does not mean that his account is only of interest to historians. On the contrary, his descriptions of recent developments in physics are thoroughly modern and may be read with profit by Americans who agree with the dictum of Henry Ford that history is bunk. The reading of the Riemann biography is not a prerequisite for understanding the monograph on topological methods. The monograph is self-contained and provides a splendidly clear explanation of such modern inventions as liquid crystals, 't Hooft monopoles, and twisted hedgehogs. The menagerie of strange objects populating the recent literature of mathematical physics is made intelligible by viewing them all as special cases of a single unifying concept. The unifying concept is the classification of spaces and mappings into discrete topological types. Each species of crystal, monopole, or hedgehog corresponds to a particular class of mappings, and the variety of species is a consequence of the inherent richness of structure allowed by the topological classification. The exploration of possible topological structures which may be relevant to physics is not yet at an end.

Monastyrsky's lucid account of the modern topological zoo can be understood by people who have no interest in Riemann, and his biography of Riemann can be understood by people ignorant of modern physics. Nevertheless, both classes of readers could profit enormously from studying that half of this book which is less familiar to them. The physicist could learn some history, and the historian could learn some physics. Both could gain a deeper understanding of their own fields by seeing them as part of a broader vision, a vision combining historical scholarship with mathematical expertise. Monastyrsky is a living bridge between the two cultures. Riemann's life and death, and the slow fruition of his ideas a hundred years later, constitute a human and intellectual drama which must be seen as a whole in order to be fully understood.

FREEMAN J. DYSON
Princeton, 1987

From the Introduction to the First Edition

This book consists of two independent parts: "Bernhard Riemann" and "Topological Methods in Contemporary Physics." They were written at different times, and I have not tried to unite them into a single entity. Therefore, I was at first surprised at Birkhäuser's suggestion that they be published under a single cover. However, after some reflection, I found this proposal remarkably appropriate: Riemann's ideas permeate contemporary mathematics and physics. Perhaps reading these two parts in conjunction will force the reader to appreciate this fact.

Riemann was one of the few great minds whose works combined the ability to solve difficult problems with a deep philosophical penetration into the basic laws of the universe. This fact is reflected in the unfinished notes in which he wrote about his plans as follows:

> The works which now mainly occupy me are
>
> 1. In a way similar to that which has already been so successfully used for algebraic functions, exponential or circular functions, elliptic and Abelian functions, introduce the imaginary into the theory of other transcendental functions. In my inaugural dissertation I have provided the most necessary general preliminary work (cf. Art. 20 of this dissertation).
>
> 2. In connection with this there are new methods of integrating partial differential equations, which I have already applied to several physical subjects.
>
> 3. My main work concerns a new conception of the known laws of nature... .[1]

Recently I again had occasion to encounter these statements of Riemann when they were quoted by the outstanding contemporary mathematician I. M. Gel'fand. Speaking at a meeting of the Moscow Mathematical Society honoring his 70th birthday, Gel'fand named Riemann among his mentors, in the true sense of that word, along with the academic advisor of his graduate-student years, academician A.N. Kolmogorov (1903–1987).

---

[1] Quoted in Klein, *Development of Mathematics* , p. 233.

Several other observations are in order. The book *Riemann* grew out of a modest article dedicated to the 150th anniversary of Riemann's birth and was published in the Soviet journal *Priroda* (*Nature*) in 1976. It was written on the recommendation of Professor B.N. Delone, a remarkable mathematician and also an authority on history. In preparing the article I discovered to my surprise that there was absolutely no detailed biography of this mathematical genius in Russian. I therefore decided that a book on Riemann, incorporating not only an account of his life and works but also an appraisal of his role in contemporary science, might be useful. I am very pleased that it is also being translated into English.

This is in no way a study in history *per se*, although I did study carefully all the published materials available to me. I hope that it does not contain serious historical and factual mistakes. Unfortunately, I was not able to study Riemann's archives. Several materials, in particular Riemann's letters to his family, have been published by E. Neuenschwander.[2] These letters enrich the picture of Riemann by providing a series of interesting details, but they do not at all change the impression one gains from reading what his friends and colleagues wrote about his life.

In the several years that have passed since the Russian edition came out, the alliance between mathematicians and physicists has been strengthened and expanded. Especially impressive results in the realm of mathematical physics have been connected with the penetration of topological ideas and the development of the method of the inverse scattering problem—a powerful method of integrating nonlinear equations. Riemann's ideas and methods have found their application here as well.

Unfortunately, it was not possible to incorporate these results into the present book without undertaking serious revisions. However, one can find something about this topic in the chapter, "Soliton Particles," in Part II and in the literature cited in the Bibliography to Part II.

There has been one more (1983) outstanding result in the realm of algebraic geometry: the proof of L. Mordell's conjecture on the finiteness of the number of rational points of an Abelian variety. (Among other things, this theorem also confirms the finiteness of the possible number of solutions of the equation $x^n + y^n = z^n$ ($n \geq 3$)—Fermat's Last Theorem.) The proof of this conjecture, obtained by the German mathematician G. Faltings, uses subtle and deep methods of contemporary algebraic geometry.[3] These works can also be considered a high point in the progress achieved over the years along the path begun by the classics of algebraic geometry, including Riemann's work.

The final result I would like to mention is associated with the unexpected application of so abstract a theorem as the Riemann–Roch theorem to the theory of coding. In using this theorem the problem of constructing codes comes down to an analysis of algebraic curves.[4] All of these results and many others obtained in recent years owe their origin to Riemann's work.

---

[2]E. Neuenschwander, *Cahier du Séminaire d'Histoire des Mathématiques*, **2** (1981): 85–131.

[3]G. Faltings, "Endlichkeitssätze für abel'sche Varietäten über Zahlkörpern," *Invent. Math.*, **73** (1983): 349–366.

[4]See the review article of V. D. Goppa, "Codes and information," *Russ. Math. Surveys*, **39** (1) (1984): 77–120.

In conclusion, I wish to express my appreciation to the staff of Birkhäuser for their exceptional kindness and competent collaboration in preparing the manuscript. In spite of all the complications born of communicating over a distance, Birkhäuser has brought this project to completion. It is a pleasure to express gratitude to K. Peters,who suggested that this book be published by Birkhäuser.

I would also like to add the following list of my colleagues and friends who helped in the final preparation of the book and thank them especially for their advice on Part II: S. S. Demidov, E.I. Kats, V. S. Kirsanov, I. M. Lifshits, S. P. Novikov, Ya. A. Smorodinskii, and K. G. Boreskov.

M. MONASTYRSKY
Moscow, 1984

Acknowledgments

I would like to emphasize the role of Ronny Wells in the preparation of the first edition. He, together with Rena Schwarze Wells and the translators James and Victoria King, put in a tremendous amount of time and energy in converting the Russian manuscript to English. I also would like to express my appreciation to the (former) editors at Birkhäuser, I. Kramer, C. MacPherson, K. Peters, and P. White for their efforts in publishing the first edition despite complicated circumstances due to long distance communication.

In preparing the second edition I have taken the advice of a number of mathematicians and physicists. I especially wish to mention F. Bogomolov, K. Boreskov, S. Coleman, S. Demidov, A. Kaidalov, V. Kirsanov, M. Kléman, O. Lavrentovich, J.H. Przyticki, S. Patterson, and C. van Weert. For the opportunity to become acquainted with Riemann's manuscripts and obtain a number of rare photographs, I am indebted to H. Rohlfing, director of the library of Göttingen University. The review of J. Gray has been very helpful. The translation of the second edition has been done by my friend Roger Cooke. This is not the first time I have had the opportunity to appreciate his competence, efficiency, and unselfish assistance. I am also very pleased to note the energetic support of E. Beschler, A. Kostant, and T. Grasso. If not for their insistence, this book would never have been published. The final stage of work on the manuscript was carried out in Cambridge, New York and Orsay, where I was a guest of Harvard University, the Courant Institute of Mathematical Sciences and the Laboratoire de Physique des Solides at Université de Paris-Sud. I am grateful to F. Bogomolov, D. & V. Bernstein, S. Cappell, H. Edwards, S. Hirson, A. Joets, S. Jones-Taft, D. Kazhdan, P. Lax, D. Mclaughlin, L. Nirenberg, R. Ribotta, and C. Taubes for their hospitality and attention. The work on this book was partially supported financially by the Russian Foundation for Basic Research Grants 96-01-01876 and 96-06-807-04.

Lastly, I wish to express unfortunately belated gratitude to my mother, who always supported me through difficult times and rejoiced in my successes. She helped me a great deal in the preparation of the first edition, but unfortunately did not live to see the publication of the second. I dedicate this book to her fond memory.

December, 1998
Harvard University
Cambridge, MA

M. MONASTYRSKY

Part I

Bernhard Riemann

Chapter 1

Beginnings

"Mathematicians are born, not made."
HENRI POINCARÉ

IN the ranks of the outstanding mathematicians of the nineteenth century the name of Bernhard Riemann occupies an illustrious place. Although his career lasted only fifteen years, he made an enormous contribution to almost all areas of mathematics. He worked on the theory of the integral, the theory of functions of a complex variable, geometry, calculus of variations, the theory of electricity, and other subjects. While this short list attests to his multifaceted talents, it does not begin to suggest the power of his mind and his striking originality. Practically all of Riemann's articles contain completely new ideas, such as the introduction of the concept of Riemann surfaces, which provided the foundation for contemporary complex analysis and topology. The value of Riemann's work is significantly greater than the numerous concrete results he achieved; his exceptionally fruitful ideas stimulated further developments not only in mathematics but also in mechanics, physics, and the philosophy of the natural sciences as a whole.

Georg Friedrich Bernhard Riemann was born on September 17, 1826, in the village of Breselenz, near the city of Dannenberg in the kingdom of Hannover. His father, Friedrich Bernhard Riemann, was a Lutheran pastor who participated as a lieutenant in the Napoleonic Wars of 1812–1814, the "wars of liberation." He was in the army of the Austrian general Count Ludwig Wallmoden (1769–1862), which gained distinction in the siege of Hamburg. The army, which combined Russian, Prussian, and other allied troops, smashed the units of Marshal Davout at Mecklenburg.

Friedrich Bernhard was already middle-aged when he married Charlotte Ebell, the daughter of a court councillor. Bernhard was the second of their two boys and four girls. As a boy his health was poor; in general illnesses and premature deaths haunted all the members of his family. His mother died when he was twenty, and his brother and three of his sisters also died quite young. Riemann was always very attached to his family and maintained the closest contact with its members throughout his life. Timid and reserved by nature, he felt at ease and free in the company of his relatives.

GEORG FRIEDRICH BERNHARD RIEMANN

When he was five years old, history, especially the history of Poland, interested him most. His special interest in the history of Poland might seem a little puzzling to a modern reader, but if one looks at the events of 1830 a lot will become much clearer. The famous Polish Uprising began in November 1830 with the attack on the Belvedere Palace, the resistance of the great prince Konstantin (brother of Russian Tsar Nicolai I and the governor general of Poland). This revolt became known in history as the November Insurrection. The war lasted from January until September of 1831; the superior Russian forces finally defeated the Polish insurrects. More than 6000 leaders of the uprising were forced into exile. The exodus of elite Poles is known in history as the Great Emigration. Most of them emigrated to France but some settled in other countries, in particular, Germany. Newspapers of that time devoted a lot of space to the uprising. Not surprisingly, young Riemann was intrigued by Polish events.

An interest in history and in the humanities in general is characteristic of many great mathematicians. One has only to think of Carl Friedrich Gauss (1777–1855) who, as a student, wavered between philology and mathematics as his specialty, and Carl Gustav Jacob Jacobi (1804–1851), who participated in a seminar on ancient languages.

Soon the family began to notice his striking ability to make calculations. At the age of six, under the tutelage of his father, who was quite an educated man, he solved arithmetic problems. When he was ten, a teacher named Schulz began to work with him, but the pupil soon outstripped his master. At the age of fourteen Riemann entered directly into the third (senior) class of the Gymnasium in Hannover; after two years he transferred to the gymnasium of the city of Lüneburg, where he continued to study until he was nineteen. Riemann was not a brilliant student, although he made a serious study of such subjects of the classical gymnasium curriculum as Hebrew and theology.

Schmalfuss, the director of the gymnasium, noticed the boy's mathematical talents and allowed him to use his personal library. On one occasion he gave Riemann a textbook on the theory of numbers written by Adrien-Marie Legendre (1752–1833). Riemann studied this book, nearly 900 pages long, for six days. Various aspects of what he learned there were used some years later in his own work on the theory of numbers.

In 1846, in accordance with his father's wishes, Riemann matriculated at Göttingen University in the faculty of theology. His interest in mathematics was so strong, however, that he asked his father to allow him to transfer to the faculty of philosophy. At this time the faculty included such well-known scholars as the astronomer Carl Wolfgang Benjamin Goldschmidt (1807–1851), who lectured on terrestrial magnetism, the mathematician Moritz Stern (1807–1894), who lectured on numerical methods and definite integrals, and the "prince of mathematicians," Carl Friedrich Gauss. At the time, Gauss, who was at the height of his powers, gave a brief course on the method of least squares. Given Gauss' extremely unsociable character and his secluded way of life, it is doubtful that Riemann had any personal contact with him at this time. Stern, however, did notice Riemann's ability; he subsequently recalled of Riemann that "he already sang like a canary."[1]

By the mid 1840's, Göttingen University, where Riemann was to work almost all of his life, had already existed for more than 100 years. It was considered one of the most illustrious universities in the German kingdoms. Founded in 1734 by the British King George II, who was simultaneously the Elector of Hannover, the university was named in his honor "Georgia Augusta" and opened in 1737. George II intended it to be the best in Germany, and first-rate scholars were invited to it. Stormy periods of German history (the Seven Years' War, the Napoleonic Wars, etc.) left it relatively untouched. However, in 1837, there occurred an event that led to a sharp decline in the academic standard of the university. The new king Ernst Augustus II (the former Duke of Cumberland) abolished the democratic constitution of 1833, which had been adopted in Hannover after serious popular unrest aroused by the July Revolution of

[1] Klein, F. *Development of Mathematics*, p. 233.

1830 in France. The constitution had proclaimed the establishment of a bicameral parliament with the participation of all levels of the population, (including the peasantry), freedom of speech, freedom of the press, open legal proceedings, etc. The 1833 constitution was replaced by a new one that, in essence, reinstated the old Hanoverian constitution of 1819. All government employees—professors included—were obliged to swear allegiance to the new constitution. Despite the great dissatisfaction with the change, all swore allegiance except for seven professors who have come to be known in history as the "Göttingen Seven": the jurist Wilhelm Eduard Albrecht (1800–1876), the historians Georg Gottfried Gervinus (1805–1871, author of a history of German poetry), Friedrich Christoph Dahlmann (1785–1860), the Arabist and Hebraist Georg Heinrich August Ewald (1803–1875), the physicist Wilhelm Eduard Weber (1804–1891), and the philologist brothers Jacob (1785–1863) and Wilhelm Grimm (1786–1859), founders of classical German philology and collectors of fairy tales.

All were forced to leave the university, and Jacob Grimm, Dahlmann and Gervinus were exiled from Hannover as well. The names of these scholars were well-known, not only in all the German states but also throughout Europe. Their expulsion from the university deeply agitated society both within the country and beyond its borders. Even the governments of Prussia and Austria, which supported the king, disapproved of these extremely harsh actions. The expelled professors received posts in other universities while Göttingen University immediately lost its reputation as the best German university. The university began to revive only in 1848, when the king, who was frightened by the revolutions of that year, agreed to restore the 1833 constitution. Wilhelm Weber was among others who then returned to the university.

Ironically, this episode was to be repeated almost verbatim a century later. The coming of the Nazis to power led to the exclusion, on the basis of the racial laws of the Reich (*Reichgesetze*), of seven professors of non-Aryan origins: M. Born, (1882–1970), R. Courant (1888–1972), E. Noether (1882–1935), and others. The Nobel laureate James Franck, (1882–1964) left Germany in protest against the racial laws. Göttingen University was dealt a blow from which it has not yet recovered. After studying for one year at Göttingen, Riemann moved to Berlin. At that time such brilliant mathematicians as Carl Gustav Jacob Jacobi (1804–1851), Jakob Steiner (1796–1863), Peter Gustav Lejeune-Dirichlet (1805–1859), and Ferdinand Gotthold Max Eisenstein (1823–1852) were teaching at the University of Berlin. In terms of the quality of its mathematics faculty, the University of Berlin was better than Göttingen. It was founded by Wilhelm von Humboldt (1767–1835) in 1810, at a time when Napoleon's army was occupying the Prussian capital (those were idyllic days when a university could be opened in an occupied city). Humboldt's principles, which became the foundation of a contemporary university education, played a critical role in promoting scholarship in Germany. In contrast to a tendency traceable to Gottfried Wilhelm Leibniz (1646–1716), Humboldt believed that universities were the proper place for scholarship. "Solitude and freedom are the principles prevailing in her realm." As far as teaching went, he advocated the principle, "Lehrfreiheit, Lernfreiheit" (freedom to teach, freedom to learn). Each professor could teach any course

GOTTFRIED WILHELM LEIBNIZ

he chose, and a student could choose lectures that interested him. The only (albeit fairly strict) control over academia was the state examination that a student took upon completion of the university.

An interesting feature of German universities was the freedom with which students were able to move from one university to another. As Wilhelm Ostwald (1853–1932) noted in his *Klassiker*, "If a specialty of interest to a student is presented at a university by an eminent has-been, then the student transfers to a different university where he will be able to take this subject from a young and progressive docent."

The two years in Berlin were exceptionally important in Riemann's scientific preparation. It was here that he became friends with the excellent mathematician Eisenstein, with whom he discussed the possibility of introducing complex variables when investigating elliptic functions.

Eisenstein completed several outstanding works on the theory of elliptic functions (Eisenstein's θ-series expansion) and on the theory of invariants. He was highly esteemed by Gauss, to whom is ascribed the following statement: "There have been only three epoch-making mathematicians: Archimedes, Newton, and Eisenstein." But

CARL GUSTAV JACOB JACOBI

if we are to believe Felix Klein (1849–1925):

> Eisenstein was too much the formula-man, who, starting from computation, found in it the roots of his knowledge and was not able to grasp Riemann's general ideas on functions of complex variables, which, according to Dedekind (1831–1916), Riemann first worked out in detail in the fall of 1847, at the age of 21. In any case Eisenstein was the only mathematician with whom Riemann associated at that time. Unfortunately Eisenstein died at the age of 29.[2]

The lectures of Dirichlet proved to have the greatest influence on Riemann. Klein, in his *Development of Mathematics in the 19th Century*, wrote:

> Riemann was bound to Dirichlet by the strong inner sympathy of a like mode of thought. Dirichlet loved to make things clear to himself on an intuitive level; along with this he would give acute, logical analyses of foundational questions and would avoid long computations as much as possible. His manner suited Riemann, who adopted it and worked according to Dirichlet's methods.[3]

---

[2] *Ibid.*, p. 235.

[3] *Ibid.*, pp. 234–235.

Riemann's stay in Berlin coincided with the March Revolution in Prussia but, as far as one can judge from the literature, political events held little interest for him. It is known only that at the very height of the revolution, together with other students, he defended the palace of the Prussian king, Friedrich IV, for almost twenty-four hours. When it became clear that there was no direct threat to the king's life, he returned to the university.

In 1849 he returned to Göttingen, where he attended the lectures of Wilhelm Weber. Weber, a close friend and assistant of Gauss, was the author of well-known works on electrodynamics. He became famous for the invention (together with Gauss) of an electromagnetic method of telegraphy. Weber himself built the receiving and sending apparatus and conducted a demonstration of them in operation (one apparatus was set up in the observatory, the other in the physics institute at Göttingen University). As a person Weber was exceptionally polite and pleasant.

Riemann's acquaintance with Weber was of decisive significance for the young man's entire subsequent fate, both personal and professional. For over a year and a half Riemann worked as an assistant in Weber's laboratory. Undoubtedly, Riemann's interest in physics took shape under Weber's influence; Riemann's works, however, bear the imprint of his unique personality.

In 1850 a seminar on mathematical physics opened at Göttingen University. Riemann became a participant in this seminar, run by Weber, Johann Benedict Listing (1808–1882), Stern, and Georg Ulrich. The content of the seminar was quite varied and included investigations not only of physical and mathematical works but also of philosophical ones. Riemann's interest in the philosophy of Johann Friedrich Herbart (1776–1841) dates from this period. The views of Herbart, who had been a professor at Göttingen University since 1833, proved to have a significant influence on the development of Riemann's own scientific world view.

A student of J. Fichte (1762–1814), Herbart was an interesting figure in philosophy and left his imprint both in pedagogy and psychology. If his system of pedagogy (based on a strict regimentation of the upbringing of children) now seems archaic, his views on psychology seem, by contrast, rather prophetic. This can be seen in his work, "On the possibility and necessity of applying mathematics to psychology," which he read to the Royal Scientific Society in Königsberg on April 18, 1822:

> In the soil of speculation many things develop that do not originate out of mathematics and do not concern themselves with mathematics. I would not consider everything growing in such fashion as a weed—many noble plants can develop thus, but not one of them can achieve full maturity without mathematics.

In psychology Herbart introduced an important scientific concept, "the threshold of consciousness." His definition of philosophy is no less interesting.

> All of our thoughts can be examined from two points of view: partly as the activity of our mind, partly on the basis of what those thoughts lead us to think. In the last analysis they are called concepts. Initial concepts are

> composed under the influence of surrounding circumstances and therefore bear a rather accidental character. When life pushes a man to reflection, motivating him to give an account in his own personal thoughts, he first begins to take note of all the imprecision, incoherence, incompleteness and incorrectness of his concepts. Thus, partly as a consequence of the practical necessity to eliminate a contradiction or explain some matter, partly as a result of purely theoretical interest, a desire arises in him to correct, to amplify, to tie together—in general to put his concepts into good order. In other words, he begins to feel an urge to philosophize. Thus, philosophy is the process of developing concepts.

In keeping with the spirit of Herbart, Riemann felt that the task of science was to comprehend and explain nature logically by means of precise concepts. All of his multifaceted scholarly activity was dedicated to achieving this general goal.

In 1850 Göttingen Riemann's purely mathematical interests were concentrated on the problems of functions of a complex variable. He was unusually fortunate to have returned to Göttingen after his two-year sojourn in Berlin, where he received a brilliant background in analysis. It is quite likely that nowhere else was the scholarly atmosphere so full of geometric—or more precisely, of topological—ideas as at Göttingen University. It was there, in 1847, that the first book on topology was published: *Vorstudien zur Topologie* (*Prolegomena to Topology*) in "Göttinger Studien" by J.B. Listing. Listing had begun his studies of topology under the influence of Gauss. Gauss himself had worked on this subject a great deal, as can be seen from his scholarly legacy. Riemann was well acquainted with Listing and his work but, in point of fact, aside from the basic definition and several properties of knotted curves, could get nothing from it. This is not said to reproach Listing, but only reflects the real state of the branch of mathematics that Gottfried Wilhelm Leibniz called "analysis situs" (analysis of position). In his book *Characteristica Geometrica* (1679) he tried (in modern terms) to study the properties of figures associated with their topological rather than their metric parameters. He wrote that, in addition to the coordinate representation of figures: "We need a different analysis, purely geometric or linear, which also defines the position (situs) as algebra defines quantity." In the subsequent 150 years, except for the famous formula on polyhedra ascribed to Leonhard Euler (1707–1783) ($V - E + F = 2$, where V is the number of vertices, E is the number of edges, and F is the number of faces) and the solution of the problem of the seven bridges of Königsberg (see Fig. 8.1, Part II), nothing more appeared in the theory of "analysis situs."

The term "topology," introduced by Listing, became attached to this branch of mathematics only at the beginning of our century; Riemann used exclusively the term "analysis situs."

Riemann's first scholarly successes are associated with the introduction of topological methods into the theory of functions of a complex variable.

Chapter 2

Doctoral Dissertation

"The shortest path between two truths in the real domain lies in the complex domain."

J. HADAMARD

AT the end of November 1851 Riemann presented his doctoral dissertation, "Grundlagen für eine allgemeine Theorie der Funktionen einer veränderlichen complexen Grösse" (Foundations of a general theory of functions of a complex variable). In order to appreciate the results obtained in the dissertation, it is useful to summarize briefly the results obtained up to that time, although it must be remembered that it is, unfortunately, very difficult to establish a clear notion of what Riemann himself really knew. A general feature of scholarly articles of that era is an extremely meager citation of sources, even in those cases when the borrowing is obvious and is not, strictly speaking, being hidden. For example, in Riemann's work devoted to Abelian functions there is not a single reference to Niels Henrik Abel (1802–1829) and only one to Jacobi. In the case of Riemann's dissertation his most significant predecessors in developing the theory of functions of a complex variable were Augustin-Louis Cauchy (1789–1857) and Karl Weierstrass (1815–1897). The fact that there were no references to the work of Weierstrass in the dissertation is not surprising: his main work had not yet been published, although it was known that it had been done. The situation with respect to Cauchy is different: by 1850 Cauchy had already published many papers on the theory of a complex variable. His first significant work, "Mémoire sur la théorie des intégrales définies," (Memoir on the theory of definite integrals) was communicated to the Paris Academy in 1814, but was published only in 1827 with several emendations reflecting the evolution of his own views. In essence, he solved the following problem: *Under what conditions on a complex function $f(z)$ is the integral $\int f(z)\,dz$ along a closed contour l equal to zero?* In the main body of the text he does not discuss complex functions explicitly; rather, using a pair of real functions $P(x, y)$ and $Q(x, y)$, he obtains the following result: *The integral $\int f(z)\,dz$ is independent of the path of integration if the following* *conditions are satisfied:*

$$\frac{\partial P}{\partial x} = \frac{\partial Q}{\partial y}; \quad \frac{\partial P}{\partial y} = -\frac{\partial Q}{\partial x}. \tag{2.1}$$

Conditions (2.1) are the criterion for analyticity (holomorphy) of functions of a complex variable. In modern literature they are called the Cauchy–Riemann conditions, although they appeared in the work of Euler on hydrodynamics and even earlier in the work of Jean le Rond d'Alembert (1717–1783). The construction of the theory of analytic functions on the basis of the conditions (2.1) is the achievement of Cauchy and Riemann.

In 1825 Cauchy wrote a memoir on definite integrals with imaginary limits. It was found among his posthumous papers and was not published until 1874. In it Cauchy defined the integral of a complex function $f(z)$ in a complex domain, including an investigation of the extremely important case when the function $f(z)$ has a singularity. For such functions the integral $\int_a^b f(z)\,dz$ may depend on the path. The same work contains the first appearance of the concept of the residue of an analytic function. Recall that the residue of an analytic function $f(z)$ is defined to be the coefficient of $(z - z_0)^{-1}$ in the expansion of $f(z)$ in the series

$$f(z) = \sum_{n=-\infty}^{+\infty} a_n (z - z_0)^n.$$

In a work completed in 1831 and published in 1836 Cauchy obtained the power series expansion for an analytic function $f(z)$ and the integral representation of $f(z)$ inside a circle. Finally, his famous theorem on residues—the Cauchy residue theorem—appeared in 1846: *Let $f(z)$ be an analytic function in the region Ω having poles at the points $z_1, \ldots, z_n$; then*

$$\int_l f(z)\,dz = 2\pi i \sum \text{Res}\, f(z_i), \tag{2.2}$$

where the sum on the right-hand side extends over all the singularities of the function f inside the contour l. In all of this circle of ideas only poles were considered, that is, singularities of the type $(z - z_0)^{-n}$.

In an 1846 paper Cauchy investigated multivalued functions (for example, integrals of the type $\int \sqrt[n]{z}\,dz$) and showed that in these cases the integrals depend on the path of integration and that no simple expression like the residue formula exists. His work on multivalued functions has a preliminary character and is extremely vague. The most prominent of the French mathematicians who investigated multivalued functions was Victor Alexandre Puiseux (1820–1883), who published a long memoir in the *Journal de Mathématiques Pures et Appliquées*, published by J. Liouville (1809–1882). He studied the algebraic equation $f(u, z) = 0$, where f is a polynomial in u and z. In this article he clarified the difference between poles and branch points ($\sqrt[n]{z}$) and introduced the concept of an essential singularity, that is, a point where the Laurent series expansion contains an infinite number of negative terms (for example, the

AUGUSTIN-LOUIS CAUCHY

function $e^{1/z}$ at $z = 0$). These ideas were independently investigated by Weierstrass. Puiseux carefully examined the situation that arises when branch points occur and showed in particular that the series expansion in a neighborhood of a branch point contains fractional powers. These papers are the most significant results published before Riemann's dissertation.

Did Riemann know Cauchy's work? E.T. Bell (1883–1960), a well-known historian of mathematics, recounts in his book *Men of Mathematics* a story of the famous British mathematician James Joseph Sylvester (1814–1897). While in Germany in the 1890s Sylvester made the acquaintance of a former (unnamed) student of Riemann, who told him the following story: Having seen in the library at Göttingen University the latest issues of *Comptes Rendus* with Cauchy's papers, Riemann disappeared with them for several weeks; when he reappeared, he said, "This is the beginning of new mathematics." It is difficult to vouch for the authenticity of this (undocumented) story. However, convincing evidence that Riemann knew of Cauchy's work is provided by a list of the books borrowed by students in the library of Göttingen University during

JAMES JOSEPH SYLVESTER

the years 1846–1847.[1] Riemann was an avid reader and among the books that he borrowed were many of Cauchy's works.

This historical excursion has been undertaken only for the sake of a clearer appreciation of Riemann's originality and the significance of his results. To appreciate any scholarly work, especially if it is of exceptional quality, it is of interest to evaluate the scholarly advance made by the author. For this purpose, it is necessary to determine what information he had at his disposal.

Riemann's dissertation consists of three parts. The first part is devoted to the investigation of geometric properties of analytic functions. Riemann gives a geometric interpretation of the concept of analyticity. For this purpose he introduces two complex planes simultaneously—the domain of the argument z and the domain of the

[1] See the paper of Neuenschwander, E., "Über die Wechselwirkungen zwischen der französischen Schule und Riemann und Weierstrass. Eine Übersicht mit zwei Quellenstudien," *Archive for History of Exact Sciences* **24** (1981), 221–225, where a facsimile of one page of this list can be found.

value w of the function. One can then formulate the condition for analyticity as follows: Choose any point z_0 in the z-plane and let its image be $w_0 = f(z_0)$. Consider two neighborhoods of the points z_0 and w_0, $U(z_0)$ and $V(w_0)$, sufficiently small that the function $f(z)$ maps the neighborhood U onto V in a one-to-one manner. Now take two curves Y_1 and Y_2 in U that pass through z_0 and also take the two curves $\widetilde{Y}_1$ and $\widetilde{Y}_2$ through w_0 that correspond under f to Y_1 and Y_2 respectively. Denote the angle between Y_1 and Y_2 and the angle between $\widetilde{Y}_1$ and $\widetilde{Y}_2$ by ψ and $\widetilde{\psi}$ respectively. We shall consider these angles to be oriented, that is, we are considering the angle of rotation from Y_1 to Y_2 and from $\widetilde{Y}_1$ to $\widetilde{Y}_2$.

A mapping of a neighborhood of a point z that preserves angles and orientation is called conformal. The following proposition is generally true: *The condition for analyticity (holomorphy) of a function is equivalent to the condition for conformality.* This assertion fails only at points z where the derivative $f'(z)$ is zero. As Riemann himself observed, conformal mappings had previously been studied by Gauss in his "General solution of the problem of mapping a given part of a surface in such a way that on the infinitesimal level the image is similar to the region mapped," printed in 1822 in *Astronomische Abhandlungen*. Gauss had encountered this problem in connection with his investigations in geodesy. It arises every time one attempts to map from a globe to an ordinary geographic map. It is noteworthy that Joseph-Louis Lagrange (1736–1813) obtained the Cauchy–Riemann conditions in 1779, also in connection with the solution to a cartographic problem. He considered the global problem of conformally mapping a plane onto itself and obtained, naturally, the same equations. If Lagrange had realized that any surface is locally similar to a plane, he would have obtained at once the condition for conformal equivalence of two surfaces.

Riemann began his work where Gauss left off. He based his study of analytic functions on the property of conformality, and this allowed him to avoid the use of explicit analytic formulas. The property of conformality was known already to Gauss. Gauss' *Astronomische Abhandlungen* is the only reference cited in Riemann's dissertation, and it provided the initial premises from which he began his own investigations. Riemann's basic task was to consider the behavior of an analytic function, not on the plane, but rather (as Riemann himself put it) "...on a surface spread over the plane." In his general examination of this question, Riemann turned to *analysis situs*. The basic idea is that the behavior of an analytic function on any surface can be reduced to a study of a function on a simply connected domain and the determination of the jumps of the function at branch cuts. For this purpose he conducted a thorough but essentially topological investigation that led to the first results on Riemann surfaces. We shall examine them below in greater detail.

The *connectivity* of a surface is defined by using a system of cuts. It is, in fact, a combinatorial definition—the germ of the future theory of homology. A surface is called *simply connected* if any cut divides it into separate parts. If this is not the case, the surface is said to be multiconnected. Riemann's work considers primarily surfaces with a boundary. He showed that the connectivity of a surface is independent of the system of cuts made. Specifically, if the number of cuts of a certain system equals n, and the number of simply connected pieces is m, then the difference $n - m$ does not

JOSEPH-LOUIS LAGRANGE

depend on the system of cuts and will be one and the same for the given figure. This number he called the *order of connectivity*. For example, in the case of a disk, it equals -1; in the case of a ring, 0. His definition of the degree of connectivity differs from the modern definition by 1. Riemann also applied the concept of connectivity to a two-dimensional surface and established a relation between the order of connectivity of the surface and the order of connectivity of the boundary. The number of separate boundary curves of an n-connected surface is either equal to n or less than it by an even number.

After these preparations Riemann turned to the basic problem: *Determine the behavior of an analytic function having given singularities on a multiconnected surface.* He solved it first for a simply connected domain and then, using a system of cuts and computing the jumps of the function at the cuts, reduced the general case to this special one.

In the case of a simply connected domain, he proved two basic theorems. The first theorem states that if a function $u(x, y)$ satisfies the equation of Pierre-Simon

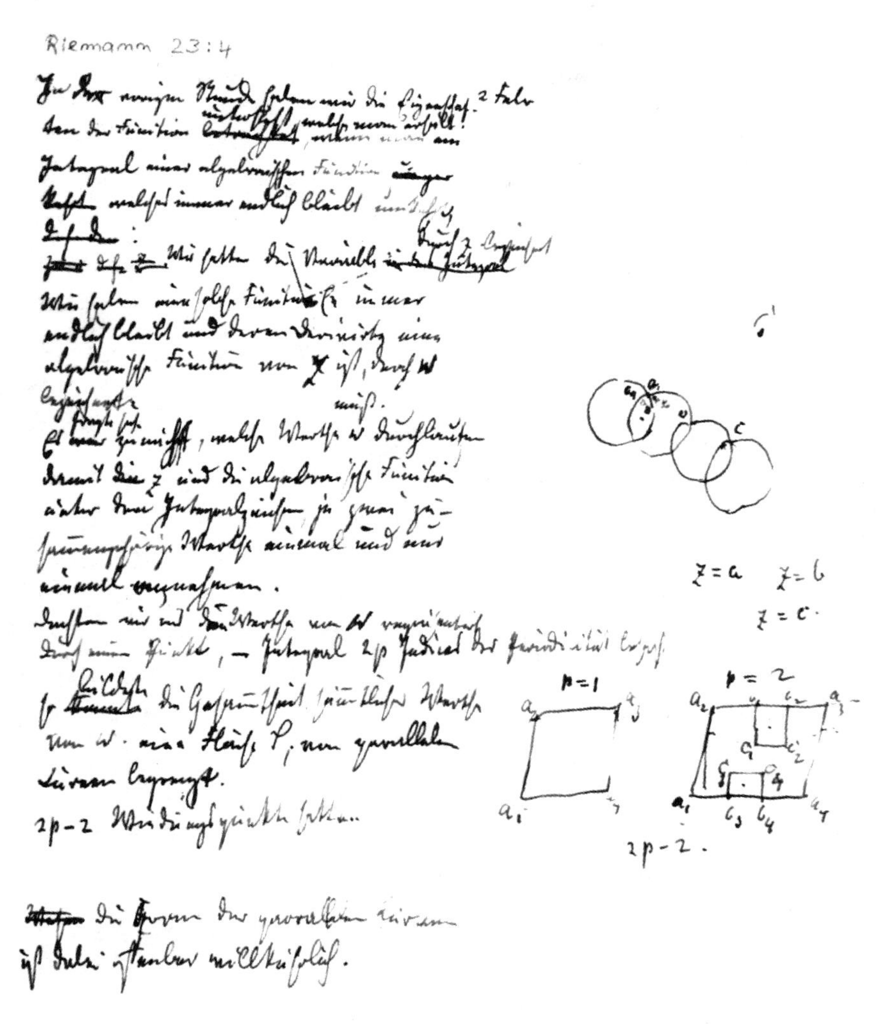

Figure 2.1: A page from the draft of Riemann's lectures on algebraic differentials (Vorlesungen uber die allgemeine theorie der integrale algebraischer Differentialien, 1861–1862). Here he considered the problem of the zeros of theta-functions. *Courtesy of the archives of Göttingen University.*

Laplace (1749–1827) in a domain, that is, if the function is harmonic,

$$\Delta u = \frac{\partial^2 u}{\partial x^2} + \frac{\partial^2 u}{\partial y^2} = 0,$$

then the function u has derivatives of all orders and, moreover, is the real part of an

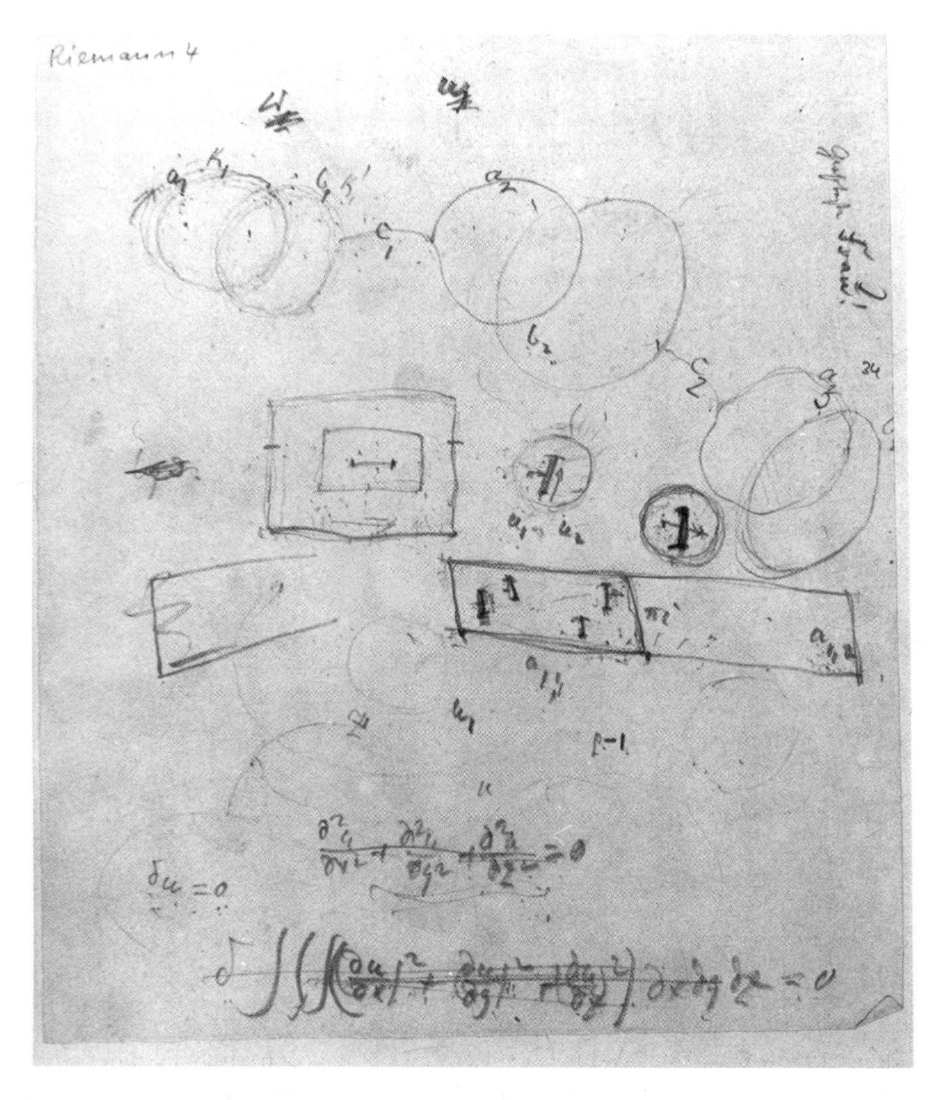

Figure 2.2: A page from the draft of Riemann's manuscript "Theorie der Abel'schen Functionen, 1857. *Courtesy of the archives of Göttingen University.*

analytic function $f(z)$. If these conditions are met, the function f is determined by u uniquely up to the addition of a purely imaginary constant term.

The second theorem—the existence theorem—says that inside a simply connected domain (it suffices to consider the case of the disk) there exists one and only one function u that is continuous up to and including the boundary, has given boundary values, and satisfies the Laplace equation inside the domain.

PIERRE-SIMON LAPLACE

For a proof of this second theorem, Riemann used a variational principle. Consider the integral in the disk

$$\iint \left(\frac{\partial u}{\partial x}\right)^2 + \left(\frac{\partial u}{\partial y}\right)^2 dx\,dy, \tag{2.3}$$

where the following conditions are placed on the function u: u is continuous up to the boundary, where it takes on the boundary value u_0, and inside the domain the integral (2.3) is finite. Consider the class of all functions u satisfying these conditions. Since the integral (2.3) is everywhere positive, it follows that there is a lower bound on its values. Suppose this bound is attained at a certain function $u = \tilde{u}$ having the given boundary value u_0. This means that the integral (2.3) attains its minimum when $u = \tilde{u}$. The variation of the integral (2.3) must be equal to zero for a minimum:

$$\delta \iint \left(\frac{\partial u}{\partial x}\right)^2 + \left(\frac{\partial u}{\partial y}\right)^2 dx\,dy = 0.$$

This condition is equivalent to the equation $\Delta u = 0$. Thus, assuming that a solution of the variational problem exists, we immediately find a suitable harmonic function, and consequently, an analytic function.

Riemann made extensive use of this method of solution, which he later called the "Dirichlet principle." He first became acquainted with it through the lectures of Dirichlet, although the method was known already to Gauss, George Green (1793–1841), and William Thomson (Lord Kelvin, 1824–1907). The Dirichlet principle contains dangerous pitfalls that are not immediately obvious.

One cannot assert a priori that when a variational integral has a lower or upper bound, there exists an actual function where it attains that extremum. Take, for example, the following problem from the calculus of variations: *Find a curve of the shortest length among all smooth curves (having continuous curvature) connecting two points A and B and passing through a third point C, assumed to be noncollinear with A and B.* It is easy to see that the length of the broken line ACB will be the greatest lower bound for the lengths of the curves under consideration. On the other hand, it is obvious that the polygonal line ACB has a corner at the point C and therefore does not belong to the class of curves having continuous curvature.

The debate associated with the Dirichlet principle played an important role in the history of mathematics and also to some extent in Riemann's life, and we will return to it later. We have not yet finished with his doctoral dissertation, which contained several results of the first order of importance, such as the Laurent series expansion for functions having poles and branch points. Probably the best-known result is Riemann's theorem on conformal mapping of simply connected domains: *Any simply connected domain of the complex plane having at least two boundary points can be conformally mapped onto the unit disk.* This is the most difficult part of the general theory of conformal equivalence of simply connected domains. Two other cases occur in the theorem: 1) the complex plane and 2) the complex plane augmented by an infinitely distant point (the extended complex plane). The interpretation of the extended complex plane in the form of a sphere (the Riemann sphere) was obtained later and was cited in his lectures.

Thus we see that Riemann's dissertation follows Gauss' geometric ideas and parallels the works of mathematicians of the French school. It provides a construction of the theory of functions of a complex variable which is significantly more complete than Cauchy's and takes a more general perspective. The concluding remarks in the dissertation show that the general nature of the problem of analytic functions on arbitrary multiconnected domains was already clear to Riemann. Moreover—and this is especially noteworthy—all these results arise as part of a general picture. He wrote,

> The reason and the immediate purpose for the introduction of complex quantities into mathematics lie in the theory of uniform relations between variable quantities which are expressed by simple mathematical formulas. Using these relations in an extended sense, by giving complex values to the variable quantities involved, we discover in them a hidden harmony and regularity that would otherwise remain hidden.[2]

[2] Riemann, B. *Gesammelte Mathematische Werke*, pp. 37–38.

Gauss, who was generally rather cold toward young people, wrote a laudatory appraisal, noting that this work attested to the outstanding independence of its author and far exceeded the requirements posed for a doctoral dissertation. This is no small tribute. There is no doubt Gauss was already in possession of many of the concepts Riemann introduced. For example, Gauss' correspondence with the astronomer Friedrich Wilhelm Bessel (1784–1846), published at the end of the last century, shows that by 1811 he had obtained the formula

$$\frac{1}{2\pi i}\int_{l}\frac{dz}{z}=1,$$

where l is a circle enclosing the origin. Acting on the principle of "Nil actum reputans si quid supperesset agendum" (consider nothing complete if anything remains undone), he published almost nothing of his mathematical works, especially in the latter years, but in letters and conversations with friends he complained more than once that young people were publishing papers containing results that he had obtained years earlier.

Chapter 3

Privatdozent at Göttingen

"To see a world in a grain of sand
And heaven in a wild flower,
Hold infinity in the palm of your hand
And eternity in an hour."

WILLIAM BLAKE

ADVANCEMENT up the career ladder at German universities was a complex and lengthy process, and obtaining a doctoral degree was only one of the mandatory steps. The system of acquiring degrees and titles influenced in no small degree the development of science, and we will look at it in somewhat greater detail. Nineteenth-century German science strongly favored candidates who had completed a university education rather than those who had graduated from the technical institutes (*Hochschulen*). Only those who had graduated from a university could occupy a teaching position and later become a professor. After obtaining a doctoral degree, one could aspire to the post of lecturer (*Privatdozent*). In order to do this it was necessary to present to the university council a *Habilitationsschrift*, a competitive composition including original work, which was presented as a small course of lectures on any special branch of knowledge. In addition, one had to deliver a *Habilitationsvortrag*, a probationary report on a narrow topic chosen by the university council. After this, if the outcome was favorable, one obtained the right to teach at the university.

A *Privatdozent* received no salary from the university but was paid fees by the students who attended his course. One could remain a lecturer for an indefinite period; everything depended on a lecturer's success and (theoretically) his talent. As soon as a post opened for an assistant professorship at any university, a lecturer could compete for the post. No special examination was required at this point. The pinnacle of a career was the title of full professor occupying a chair. Both assistant and full professors received state salaries, but there was a significant difference in both salary and status between the two levels.

Although such a system of academic distinctions had a few drawbacks, it also had obvious merits, such as the presence of a large number of highly qualified specialists (the well-known American expert on university education, J. Hart, called lecturers the life blood of German universities). The relatively small number of professorial vacancies insured a high level of teaching and—what is especially important—a certain geographical uniformity in the distribution of the ablest scholars. Of course, for each specialty there were especially desirable posts. The University of Berlin almost always ranked first, but in comparison with Paris and St. Petersburg, Berlin exerted comparatively little centripetal force.

In December of 1853, after two years of intense work in the most diverse fields of mathematics, Riemann presented his *Habilitationsschrift* "Über die Darstellbarkeit einer Function durch eine trigonometrische Reihe," ("On the representability of a function by a trigonometric series") to the Council of Göttingen University. It was printed only after his death, in the 1867 volume of the *Abhandlungen der Königlichen Gesellschaft der Wissenschaften zu Göttingen*. It is a masterfully written work with a long historical introduction and a thorough citation of sources. (These citations were provided to Riemann by Dirichlet; see Riemann's 1852 letter to his father, p. 578 of his *Werke*.)

Trigonometric series arose simultaneously in two different areas of mathematics during the eighteenth century: in astronomy, where the movement of the planets can be naturally represented by means of periodic functions, and in partial differential equations, in the problem of the vibrating string. Although the two areas sometimes present identical mathematical problems, there seem to be no parallels in the works of the most eminent mathematicians in these two fields. Astronomers represented any function they needed by trigonometric series, without much discussion, while specialists on differential equations made this representation a topic of heated debate. Even when one and the same author worked in both fields, for example, Euler or d'Alembert, the approach to trigonometric series in astronomical investigations was completely different from that used to discuss the vibrating string. The solution of the equation of the string

$$\frac{1}{c^2}\frac{\partial^2 u}{\partial t^2} = \frac{\partial^2 u}{\partial x^2} \tag{3.1}$$

gave rise to a level of polemic never before seen in the history of mathematics. D'Alembert, Euler, Daniel Bernoulli (1700–1782), and Lagrange were deeply involved in the dispute.

In 1746 d'Alembert published his first work on the solution of Eq. (3.1) for a string with fixed ends in the form $u(x,t) = \phi(x - ct) + \psi(x + ct)$. It gave rise to other questions: *What class of functions can be a solution of the equation of the vibrating string? Can one represent these functions as trigonometric series?*, etc. The argument, in which any two participants had diametrically opposed points of view, raged for thirty years. Despite the passions so aroused, the debate was in no way resolved and, in fact, could not have been settled at the time. The question of representing functions by a trigonometric series is a difficult problem in analysis and could not be solved in the eighteenth century, when the very concepts of continuity and differentiability were far from clear.

The next fundamental step was made by the French mathematician Jean-Baptiste

JEAN-BAPTISTE JOSEPH FOURIER

Joseph Fourier (1768–1830). In his classic work *The Analytic Theory of Heat* (1822) he obtained a series expansion for an arbitrary function f (Fourier series):

$$f(x) = \sum a_n \sin nx + b_n \cos nx,$$

where

$$a_n = \frac{1}{\pi}\int_{-\pi}^{\pi} f(x)\sin nx\,dx, \quad b_n = \frac{1}{\pi}\int_{-\pi}^{\pi} f(x)\cos nx\,dx.$$

It should be noted that Fourier series had appeared earlier in the works of C. Clairaut (1713–1765) and Euler, but in the thick of the discussions the correct conclusions were not drawn from them. It is curious that Fourier's first work was presented to the Paris Academy in 1807, but was rejected by Academicians Lagrange, Laplace, and Legendre. As a sort of consolation prize it was recommended to Fourier that he develop his ideas and offer a work in the competition for the Academy's grand prize. The theory of heat was the subject of the prize in 1812. Although Fourier did in fact receive the prize, the work was nonetheless considered insufficiently rigorous and too inconclusive to be published in the Academy's *Mémoires*. Only in 1824, after Fourier had become secretary of the Academy, was he able to get his work published in the *Mémoires*.

The functions that Fourier investigated arose out of physical problems, essentially from solving the heat equation:

$$\frac{\partial u}{\partial t} = k^2\left(\frac{\partial^2 u}{\partial x^2} + \frac{\partial^2 u}{\partial y^2} + \frac{\partial^2 u}{\partial z^2}\right).$$

In particular, they were sufficiently smooth. Fourier himself did not attempt to find general conditions for the validity of the series expansion. The first paper devoted to this problem, by Dirichlet, was published in 1829 in Crelle's *Journal*, the leading German mathematics journal. Dirichlet found several conditions that were sufficient for the representation of periodic functions by a trigonometric series.

This probably is all the information at the disposal of Riemann—or science generally—when Riemann began working on the problem. Because of the chaos that reigned at the time in the fundamental questions of integrability and convergence (the concept of uniform convergence did not yet exist), and also the vagueness of the basic concepts, Riemann began his investigation by determining a necessary and sufficient condition for the existence of an integral—the condition of Riemann integrability. In particular, he introduced his celebrated example of a function having an infinite number of discontinuities that is nevertheless integrable. The Riemann function $R(x)$ is defined on the segment $[0, 1]$ by the formula:

$$R(x) = \begin{cases} \dfrac{1}{q} & \text{if } x \text{ is rational of the form } \dfrac{p}{q} \\ 0 & \text{if } x \text{ is irrational.} \end{cases}$$

This function is discontinuous at rational points and continuous at irrational ones. The second part of Riemann's *Habilitationsschrift* is devoted directly to the problem of representing functions by a trigonometric series. Riemann emphasized the distinction by the form in which he posed the problem.

> While preceding papers have shown that if a function possesses certain properties, it can be represented by a Fourier series, we pose the reverse question: *If a function can be represented by a trigonometric series, what can one say about its behavior?*[1]

In this subject Riemann not only proved several remarkable theorems, he also (more importantly) discussed the theoretical problems of the theory of trigonometric series and suggested new approaches to solving them. His ideas and methods exerted decisive influence on the whole subject of trigonometric series in the subsequent decades. Here is a brief list of the results of this section of his paper.

1. He obtained a necessary condition for a (Riemann-)integrable function to be representable by a trigonometric series:

[1] See Riemann's *Habilitationsschrift* in *Gesammelte Mathematische Werke*.

Riemann's theorem. *The Fourier coefficients of any integrable function a_n, b_n tend to zero as $n \to \infty$.*

This theorem was later extended by Henri Lebesgue (1875–1941) to the class of Lebesgue-integrable functions.

2. He constructed a method of generalized summation of trigonometric series. The following question, posed by Riemann, is fundamental for the theory of trigonometric series: *Given a trigonometric series T*

$$T(x) = a_0 + \sum_{n=1}^{\infty}(a_n \cos nx + b_n \sin nx) = A_0/2 + \sum A_n(x),$$

is $T(x)$ the Fourier series of a function $f(x)$?

In particular the series $T(x)$ may be convergent, and then it is natural to ask whether it is the Fourier series of its sum. The answer is demonstrably negative unless some restrictive hypotheses are imposed, since the sum of the series is not necessarily integrable.

3. Riemann had the exceptionally original idea of solving the following problem, which is crucial in this circle of problems: *If $T(x)$ converges and the coefficients a_n and b_n tend to zero, then the series*

$$(1/2)A_0x^2 - \sum \frac{A_n(x)}{n^2} = \Phi(x) \tag{3.2}$$

converges absolutely and uniformly to a continuous sum.

The series (3.2) is obtained from the series $T(x)$ by integrating formally twice. Georg Cantor (1845–1918) later showed that the assumption that the coefficients a_n and b_n tend to zero is a consequence of the assumption that the series converges. Leopold Kronecker (1823–1891) showed that the absolute convergence of the series (3.2) can be proved without this assumption.

Riemann's basic idea was to use the function $\Phi(x)$ and regard the series $T(x)$ as a "generalized second derivative." We denote the corresponding second difference quotient by $R_h(x)$:

$$R_h(x) = \frac{\Delta_{2h}^2 \Phi(x)}{4h^2} = A_0 + \sum_{n=1}^{\infty} A_n(x)\left(\frac{\sin nh}{nh}\right)^2 \tag{3.3}$$

where $\Delta_{2h}^2 \Phi(x) = \Phi(x+h) + \Phi(x-h) - 2\Phi(x)$.

The series (3.3) is formally equal to the series $T(X)$ when $h = 0$. Riemann's idea was to define the sum of $T(x)$ using the series (3.3) in cases when $T(x)$ diverges. If the series $T(x)$ is the Fourier series of a function $f(x)$, it is Riemann-summable almost everywhere, in particular at every point of continuity of $f(x)$. The idea of considering a generalized second derivative is worth noting from a modern point of view. The generalized second derivative may exist when the ordinary second derivative (and even the first derivative) fails to exist. By this fundamental expansion of the class of functions that could be studied Riemann at once approached the concept of distributions. It is precisely in the context of the theory of distributions, which arose in the mid-twentieth century, that it was possible to solve the fundamental problems of the theory of Fourier series and integrals.

In order to appreciate Riemann's results fully, one must remember that serious investigations on the foundations of analysis had only just begun. A naive "physical" picture that regarded each continuous function as differentiable and each bounded function as integrable was quite widespread. The example of a continuous, nowhere differentiable function, constructed by Weierstrass in 1872, caused a sensation. The Weierstrass function is also defined using a trigonometric series:

$$f(x) = \sum b^n \cos(a^n \pi x),$$

where a is an odd integer, $0 < b < 1$, and $ab > 1 + (3\pi/2)$. This series converges uniformly and defines a continuous function that is not differentiable at even one point.

Problems of trigonometric series attracted the attention of many of the major mathematicians of the late nineteenth and early twentieth centuries such as Lebesgue and Cantor. The study of problems of representing functions by Fourier series and the Fourier integral led to the discovery of unusual properties of functions and the origins of measure theory and the Lebesgue integral.

As sometimes happens in science, what at one stage does not appear even to be a question later becomes a complicated problem and still later receives an unexpected solution. Operations that are not valid in one class of functions become valid when the class of functions and the operations themselves are expanded. Thus, the naive assurance (mentioned above) that had existed in operations with functions and integrals mentioned earlier received a reliable foundation. In reality each bounded (measurable) function is integrable, not in the sense of Riemann, but in the sense of Lebesgue. Continuous functions can be differentiated when regarded as distributions (the δ-function of Paul Dirac (1902–1984) arises at this point).

For most mathematicians the development of this idea would have become an entire life's work, but other ideas were already inspiring Riemann. Several problems attracted him at once. For a long time he had wondered about the connection between electricity, light, and magnetism. The recent experiments of Rudolf Kohlrausch (1809–1858) on the change of residual charge in a Leyden jar interested him. He believed he could explain this phenomenon, but first he had to submit his *Habilitationsvortrag*. He prepared three topics for this lecture. Two were associated with his investigations on electricity, while a third was on geometry. To his surprise, at Gauss' recommendation, the council chose his work on geometry. On June 10, 1854, Riemann read his probationary lecture, "Über die Hypothesen, welche der Geometrie zu Grunde liegen" (On the hypotheses that lie at the foundations of geometry).[2]

[2]See Riemann's *Habilitationsvortrag* "Über die Hypothesen, welche der Geometrie zu Grunde liegen" in *Gesammelte Mathematische Werke*.

What is space? Does geometry have a relationship to experience? These questions bring us to a complex world where mathematics collides with physics and, more dangerously, with philosophy. The purely mathematical concept of a space of n dimensions arises in many works and is a natural topic for study in the problem of describing systems with many degrees of freedom. For example, the kinetic theory of gases, operating with a phase space of $6N$ dimensions (N is the number of molecules in a one-gram molecular mass, known as the *Avogadro number*: $N \sim 6 \cdot 10^{23}$). The problem of geodesic (shortest) lines on a multidimensional ellipsoid are examples where n-dimensional spaces arise.

It is striking that even though Lagrange had introduced the concept of four-dimensional space (where time t is added to three spatial coordinates) in the eighteenth century, it was only as a convenient means for solving problems in mechanics. The study of the geometry of n-dimensional spaces proceeded very slowly, mainly—as Felix Klein authoritatively asserts—because of "the expected objections [by philosophers]" that an n-dimensional space is nonsense.[3] The discovery of non-Euclidean geometry in 1829 by Nikolai Ivanovich Lobachevskii (1792–1856), in 1831 by János Bólyai (1802–1860), and still earlier by Gauss[4] provided still more sources of controversy and misunderstanding.

The prevailing conviction of the scientific community that Euclidean geometry reflected reality was largely based on purely intuitive philosophical notions (for example, on the very influential Kantian idea of the a priori existence of the concept of space) and was not shared by the founders of non-Euclidean geometry. However the latter understood the full complexity and relativity of any assertion of the hypothesis of the "non-Euclidean" character of physical space. There is some very inconsistent literature on this subject connected with Gauss. One can say with assurance only that the famous measurements of the angles of the triangle formed by the three peaks of Brocken, Hohehagen, and Inselberg in the Harz Mountains, which Gauss made in 1816 during the geodetic survey of the Kingdom of Hannover, were not related to any test of the "Euclidean" nature of the geometry of physical space. A glance at his famous "Disquisitiones generales..." (1827) distinctly shows his belief that the measurement of the angles of a triangle on the surface of the earth cannot exhibit any deviation from two right angles within the limits of error of measurement. On this point he referred to the results of the geodetic measurement of 1816, in which the difference of the angle sum from 180° was 14.85348 seconds of arc.[5]

The possibility of measuring angles on an astronomical scale was considered by Lobachevskii. In the first part of his foundational work "On the laws of geometry"

---

[3] Klein, F. *Development of Mathematics*, p. 156.

[4] Gauss' correspondence, published at the end of the nineteenth century, shows that, beginning roughly in 1794, he seriously worked on geometric problems and mastered many concepts later developed by Lobachevskii (1792–1856) and Bólyai (1802–1860).

[5] It is instructive to note that a direct reading of the classics of science enables the student to avoid repeating widely held but sometimes utterly false beliefs. This remark applies to the present author. In the first edition, relying on the authority of Klein, I wrote that Gauss looked for an experimental determination of the geometry that holds in nature, by finding the magnitude of the difference between the angles in Brocken–Hohenhagen–Inselberg triangle and π.

NIKOLAI IVANOVICH LOBACHEVSKII

(1829) he estimated the difference from π radians for the sum of the angles of a triangle formed by a fixed star and two points of the earth's orbit. In doing this he made use of the data on the parallax of two stars: Keida (in the constellation Eridanus) and Sirius. In both cases the deviation from π was significantly less than the possible errors of measurement. For example, in the triangle formed with the star Keida $\left(\sum_{i=1}^{3} \alpha_i\right) - \pi = 0.000372$. Having thereby shown that even on such a scale it is still impossible to exhibit a single "true" geometry, he mentioned the constellations of Andromeda, Orion, and others, to which the distances are many times greater than "the distances from our earth to the fixed stars." Lobachevskii concluded his discussion of this section with cautious but penetrating reasoning on the relations between different geometries and real space and among the various geometries themselves:

> ...The new (non-Euclidean) geometry, whose foundation has been laid here, even if it does not exist in nature, nevertheless may exist in our imagination; and, while remaining unused for actual measurements, it opens a new, extensive field for the mutual application of geometry and analysis.

In studying the work of the founders of non-Euclidean geometry, one sees how much more profound and substantive was their approach to the problem of the relation between geometry and experience than the views of the professional philosophers.

We owe our understanding of the difference between geometry as a logical structure and as a reflection of physical reality to the efforts of many of the greatest minds. In this field the name of Riemann stands on a level with those of Albert Einstein (1879–1955) and Henri Poincaré (1854–1912).

Riemann's *Habilitationsvortrag* consists of two parts: In the first he considered the purely mathematical problem of defining n-dimensional space with a given metric and studied several properties of an object that he called a *Mannigfaltigkeit* (manifold). He posed the following question: *How does one define the distance between two points of a manifold, irrespective of the space in which it lies?* This intrinsic definition of the distance gives a metric on the manifold. At the same time, he considered only those manifolds in which a sufficiently small neighborhood of a point has the metric of the ordinary Euclidean space:

$$ds^2 = \sum_{l=1}^{n} (dx^l)^2. \tag{3.4}$$

In the general case he proposed that the metric be given by a positive definite quadratic form, where the square of the element of length is

$$ds^2 = g_{jk}\, dx^j\, dx^k$$

(summation on repeating indices is assumed). Spaces in which one can introduce such a metric have come to be called "Riemannian." Riemannian spaces possess a series of remarkable properties, but we shall discuss only the properties that stimulated Riemann himself to undertake their study.

The simplest example of a Riemannian space is a surface. It was apparently Euler who first noted that a space can be described not only by the coordinates of the space in which it lies but also by introducing them directly onto the space itself. However, it was Gauss who proposed a detailed study of the intrinsic properties of the surface, that is, properties that are independent of the way in which it is imbedded in any particular space. It is interesting to note that he obtained his remarkable results on differential geometry in 1816, during a period of active practical work. On assignment by the government, he conducted thorough geodetic surveys in the kingdom of Hannover. Various aspects of this practical activity led Gauss to two of his most important achievements in pure scholarship: the method of the least squares and his paper "Disquisitiones generales circa superficies curvas," (General investigations of curved surfaces, 1827).

The extremely important concept of total (Gaussian) curvature is also due to him. Consider a sufficiently small piece (a small neighborhood) U of a surface E. At each point of U belonging to the surface E one can assign a normal l—a unit vector normal to the surface. The set of values of the vector l will be a certain domain V on the unit sphere S^2. It is called the *spherical image* of the domain U. The Gaussian curvature

CARL FRIEDRICH GAUSS

K at a point is the limit of the ratio of the area of V to the area of U as U shrinks to the point. The theorem that Gauss himself called his *theorema egregium* asserts that the curvature K does not change if the surface is bent.

By a *bending* of a space is meant a transformation that does not change the distances between points or the angles between curves that lie on the surface. Surfaces are divided into three classes, depending on the sign of the curvature K: surfaces of positive curvature (for example, a sphere), surfaces of zero curvature (a plane or a torus), and surfaces with negative curvature (for example, a doughnut with two holes or a one-sheeted hyperboloid). (See Fig. 9.2, Part II.) From the *theorema egregium* one sees immediately that a flat surface cannot be bent in such a way as to obtain the surface of a sphere, but it can be bent into a cylinder or even a cone.

Another property of the Gaussian curvature, which can also be taken as a definition, is as follows. Consider a triangle on the surface E formed by three lines of shortest length (geodesic lines). Then the integral of the curvature (with respect to the element of area of the triangle) equals

$$S = \int K\,ds = \alpha_1 + \alpha_2 + \alpha_3 - \pi,$$

where α_1, α_2, and α_3 are the angles between the geodesic lines. This second definition is more convenient for Riemann's generalization to multidimensional space.

In his work Riemann introduced the concept of curvature in a two-dimensional direction. Take any point x_0 in an n-dimensional Riemannian space M^n and consider a two-dimensional surface formed by geodesic lines, having at the point x_0 tangent vectors lying in a given plane. This surface has a curvature K called the curvature in this two-dimensional direction.

Riemann managed all this without a single formula. At the end of the first part of the work, he studied multidimensional spaces of constant curvature.

His spaces of constant curvature arise from the physical requirement that "figures" can move in such spaces without expanding or contracting. In modern terms, one can explain them in the following way. Consider any three points in the space M^n. Let us determine the distances between them, having chosen a fixed metric. If it is to be possible to move any such triangle as a rigid body in space, there must be a certain symmetry group that transforms the space and preserves the distance between points. For example, if we consider a space with the metric (3.4), we arrive at the usual Euclidean space R^n, where the curvature K equals zero. In this way, but by a more complex means, one can obtain a space of constant positive or negative curvature. It is remarkable that classical geometries are associated with each of the three types of spaces: elliptic (Riemannian) geometry with spaces of positive curvature, parabolic (Euclidean) geometry with spaces of zero curvature, and hyperbolic (Lobachevskian) geometry with spaces of negative curvature.

In the second part of his lecture, Riemann turned to a discussion of the properties of a real space that make it possible to single out, from among the infinite set of possible metrics, the one that applies to reality, that is, he essentially posed the question of the relationship between geometry and the real world. Here he examined two questions separately: *First, what is the dimension of a space? Second, which geometry describes physical space?* In discussing the first question, Riemann clearly pointed out that the notion of dimension current at the time assumed a concept of continuity. In reality, when we talk about a set consisting of a finite number of points, the only dimension that can be introduced for it is simply the number of points; but when the number of points is infinite, the situation changes. Actually, the concept of dimensionality received its full logical foundation only after the discovery of the theory of sets by Cantor. He showed that if continuity is omitted, then the concept of dimensionality loses all meaning. For example, one can establish a one-to-one correspondence between points of a straight line, a plane and (in general) a space of any finite number of dimensions. Only the additional requirement of continuity allowed L.E.J. Brouwer (1881–1966) to prove the following fundamental theorem: *There does not exist a one-to-one and continuous mapping with continuous inverse of a space M^n onto M^m (given $n \neq m$).* If we assume, however, that the real world has dimensionality, for example, three dimensions, then the question arises: *Is it bounded or not?* In other words, can one represent it the way we represent a closed two-dimensional surface in three-dimensional space (for example, a sphere), or is it an unbounded surface of the type of a plane?

The answer to this last question is highly nontrivial even for two-dimensional manifolds, and depends strongly on the sign of the Gaussian curvature of the surface. For manifolds with $K > 0$ the answer is positive. The situation with manifolds of negative curvature is much more complicated. For example, it follows from a famous theorem of David Hilbert (1862–1943) that the hyperbolic plane cannot be isometrically imbedded in three-dimensional Euclidean space as a regular complete surface.

The physical insight with which Riemann posed the question is striking in its prophetic vision.

> The question of the validity of geometry in the infinitely small is related to the question of the internal basis of metric relationships of the space. For this question, which can certainly be considered in the theory of space, the above remark is applicable, that is, that a discrete manifold contains an intrinsic metric relation, while this must be added for the continuous case. Therefore, either the reality which is the basis for the space must form a discrete manifold or the basis for the metric relation must be sought elsewhere in the connecting forces.
>
> The answer to this question can only be found in that, as experience has proven, one proceeds from phenomena, in which Newton laid the foundation, to facts that are not explicable on the basis of the previous knowledge, and which lead to a gradual reworking of the theory. The investigations that we have carried out here, and which proceed from general concepts, can only help prevent this work from being hindered by limitation of the concepts, and prevent progress in the recognition of the relationship of things from being restricted by traditional prejudices.
>
> This leads to the realm of another science, into the area of physics, which the nature of today's occasion does not allow us to enter.

Not until sixty years later, after the creation of the general theory of relativity by Einstein, was Riemann's idea of defining a metric by external gravitational forces given spectacular confirmation. In Einstein's equations, the metric properties of a space (its curvature) are determined by a gravitational field.

Riemann's *Habilitationsvortrag* is nowadays regarded as one of the most remarkable works in the history of all science, not only mathematics. Therefore, it may be interesting to trace the route over which Riemann's ideas gained acceptance.

In Riemann's audience, only Gauss was able to appreciate the depth of Riemann's thoughts. Dedekind, a well known mathematician who was not only a close friend but also Riemann's first biographer, describes Gauss' impressions: The lecture exceeded all his expectations and greatly surprised him. Returning to the faculty meeting, he

GEORG CANTOR

spoke with the greatest praise and rare enthusiasm to Wilhelm Weber about the profundity of the thoughts Riemann had presented. Nevertheless, Riemann's paper was not published during his lifetime and apparently remained unknown to the majority of major mathematicians. For example, there is no mention of it in the two inductions of Riemann into the Berlin Academy of Sciences.

A report of the lecture published by Dedekind in 1868 in the journal *Göttinger Abhandlungen*[6] aroused interest. However, it analyzed primarily the purely mathematical part of the work associated with Riemannian geometry. As a result of the work of a whole constellation of outstanding mathematicians, including E. Christoffel (1829–1900), E. Beltrami (1835–1900), L. Bianchi (1856–1928), G. Ricci (1853–1925), T. Levi-Cività (1873–1941), E. Cartan (1869–1951), and many others, Riemannian geometry became an independent science, closely associated with algebra, topology, and complex analysis. It became the mathematical foundation of the general theory of relativity.

[6]*Abhandlungen der Königlichen Gesellschaft der Wissenschaften zu Göttingen*, **13**.

In discussing this brilliant work, I would like to emphasize again that Riemann drew on concrete physical problems in the posing of his sometimes quite abstract mathematical research. Riemann developed the analytic apparatus of Riemannian geometry while solving a problem for a competition set by the eminent Paris Academy. Although not a word in this paper mentions his Göttingen lecture "Über die Hypothesen...," the epigraph that preceded the article speaks for itself: *Et his principis via sternitur ad majora* (And with these beginnings the path extends to greater things). The problem posed was as follows: *To determine what the thermal state of an arbitrary solid body must be in order that a system of curves that are isotherms at a given instant shall remain isotherms at all times*, that is, so that the temperature at a point can be expressed as a function of time and two auxiliary variables.

In view of a certain incompleteness and brevity of the proof, the work was not awarded a prize by the Academy and remained unpublished until 1876, when it appeared in his collected works, published by Heinrich Weber (1842–1913) and Dedekind. After publication of the "Paris paper," it became clear that Riemann had been in possession of machinery sufficiently advanced for the construction of what is now called Riemannian geometry. In particular, this paper contained the definition of the curvature tensor and a formula for total (Gaussian) curvature in the case of n-dimensional space and even the definition of connection coefficients, introduced independently in 1869 by Christoffel (Christoffel symbols), and their connection with a Riemannian metric in the case of a space of constant curvature.

While the part of the "Hypothesen" devoted to Riemannian geometry was accepted right away and received immediate development, especially in the works of the Italian school, Riemann's ideas on physics remained completely uncomprehended. It is likely that only H. Helmholtz (1821–1894) in 1868 and the remarkable British mathematician William Kingdon Clifford (1845–1879) in 1870 paid attention to this latter side of Riemann's work. The idea of connecting a metric with measurement by means of a system of rigid bodies was a familiar one to Helmholtz. As for Clifford, delivering a paper entitled "On the Space Theory of Matter" before the Cambridge Philosophical Society on February 21, 1870, he said:

> I hold: 1) that small portions of space *are*, in fact, of a nature analogous to little hills on a surface that is on the average flat; namely, that the ordinary laws of geometry are not valid in them; 2) that this property of being curved or distorted is constantly being passed on from one portion of space to another after the manner of a wave; 3) that this variation of the curvature of space is what really happens in the phenomenon that we call the *motion of matter*, whether ponderable or ethereal; 4) that in the physical world nothing else takes place but this variation, subject (possibly) to the law of continuity.[7]

[7] From William Kingdon Clifford, *Mathematical Papers*, Robert Tucker, ed., London: Macmillan and Co., 1882.

ARTHUR CAYLEY

It is remarkable that it was Clifford who first appreciated Riemann's idea. Clifford died at an even younger age than Riemann, but left an important imprint on science. His scholarly interests were connected with the study of multidimensional algebraic and geometric structures. In algebra he continued the investigations on hypercomplex numbers, which had been begun by W.R. Hamilton (1805–1865), H.G. Grassmann (1809–1877), and A. Cayley (1821–1895). He constructed an example of a noncommutative but associative multidimensional algebra (the Clifford algebra). For many years, his results remained important for the physics of studying elementary particles with half-integer spin (particles obeying the Fermi–Dirac statistics). In geometry he developed the work of C. von Staudt (1798–1867), Cayley, and Klein on the foundations of geometry (the relationship between non-Euclidean and projective geometries). He also obtained important results on Riemann surfaces (a model in the form of a bounded surface).

Clifford's paper in Cambridge had little influence on the subsequent development of gravitational theory. Later on, Einstein came to appreciate Clifford's work; his general theory of relativity decided the question of the nature of gravity by using the geometry of space, basing itself on Riemannian spaces of constant curvature.

In recent years, an interest in the geometry of small distances has arisen under

FELIX KLEIN

the influence of new attempts to unite quantum and gravitational phenomena; that is, the theory of a quantum fluctuation of geometry (of a metric)—geometrodynamics. The basic features of this theory, whose main proponent is the American physicist J. Wheeler, are as follows. The metric of a Riemannian space defines the structure of the space only at large distances; for small distances, on the order of the Planck scale (that is, at distances $L \sim 10^{-33}$ cm, $L = \hbar G/c$, where $\hbar$ is Planck's constant, G is the gravitational constant and c is the speed of light), the metric is not determined but instead fluctuates. Thus, we come to a consideration of the indeterminate structure of space: space appears "smooth" only at first glance. On the microscopic level its structure is pebbly—Clifford hills. Not only does the geometry of the space fluctuate, but so does its topology: If we observe closely, we ought to see not only little hills but also holes. Thus, the equation of quantum theory of gravitation is analogous to the Schrödinger equation in quantum mechanics; the amplitudes of probability of transition from one geometry to another ought to play a role. A natural question arises: *Beyond general aesthetic dissatisfaction with the gulf between gravitational theory and quantum theory, what physical phenomena force us to consider such fantastic possibilities?* These phenomena, which we can only mention in passing, connect the world of elementary particles with the whole universe.

HEINRICH WEBER

One striking consequence of the general theory of relativity is a prediction of the formation of black holes as a result of gravitational collapse. The formation of black holes, however, requires such a gigantic compression of matter (density on the order of 10^{93} g/cm^3) that quantum effects become important. The most striking discovery of recent years is the one made by the British physicist S. Hawking, who proved that a black hole is a source of stationary thermal radiation (quantum vaporization of black holes—Hawking's effect). Hawking also obtained important quantitative results in fluctuation-geometric theory (in Wheeler's picturesque phrase, the space-time theory of foam).

The preceding account has endeavored to give only a minimal idea of the influence of Riemann's work on the subsequent development of science. However, I would also like to take note of yet another feature of his work—an aspect that, in fact, leaps to the attention of the modern reader. This is the clear boundary that Riemann established between results that could be rigorously proved and hypotheses that could not be verified on the basis of the resources of the science of his day. For example, in connection with the introduction of the concept of curvature, he clearly distinguished

RICHARD DEDEKIND

the finite, but unbounded space of positive curvature from unbounded space with negative curvature. The importance of this distinction can be seen in the theory of relativity, where the existence of bounded models of Friedman type leads to collapse; in the case of open models, collapse does not occur. In contrast to the generations of philosophers who discussed the topic, "Was there a beginning, and will there be an end of the world?" Riemann realized perfectly well that the choice of an adequate model of reality could be obtained only from astronomical observations (for example, a measurement of the density of intergalactic gas) which at the time were impossible to carry out. Therefore, he wrote:

> For an understanding of Nature, questions about the infinitely large are idle questions. It is different, however, with questions about the infinitely small. Our knowledge of their causal relations depends essentially on the precision with which we succeed in tracing phenomena on the infinitesimal level.[8]

[8]Riemann, B. *Gesammelte Mathematische Werke*, p. 285.

Another of Riemann's ideas remains as yet undeveloped, namely that the continuity of space breaks down at hyper-small distances, but we already know the path along which Riemann suggests seeking an answer.

In 1854 Riemann finally obtained the right to teach. Not long before, in September, he had read a paper "On the laws of distribution of static electricity" at a session of the Göttingen Society of Scientists and Physicians. In a letter to his father Riemann recalled, among other things, that "the fact that I spoke at a scientific meeting was useful for my lectures." In October he set to work on his lectures on partial differential equations. Riemann's letters to his beloved father are full of stories of the difficulties he encountered. Although only eight students altogether attended his lectures, Riemann was completely happy. Gradually he overcame his native shyness and established a rapport with his audience.

Chapter 4

Riemann and Dirichlet

"Gentlemen, we do not have time for Gaussian rigor."
CARL JACOBI

RIEMANN'S scholarly achievements were greeted rather coolly by his colleagues and, more importantly, by the university administration. The successor of Gauss (who had died in 1855) was Riemann's old friend Dirichlet. It was Dirichlet who, with great difficulty, succeeded in obtaining a small paid post in the department for Riemann. Not until November of 1857 did Riemann obtain the position of assistant professor. The years of Riemann's collaboration with Dirichlet (1855–1859) were by far his most productive. Illness had not yet undermined his strength, and the opportunity to pursue his own investigations was all that Riemann required for complete happiness.

In 1857, the memoir "Theorie der Abel'schen Functionen" appeared in the fifty-fourth volume of the *Journal für die reine und angewandte Mathematik*. It contained work that had been done in the period 1851–1856 and expounded in lectures of 1855–1856. Although only three people attended these lectures, one of them was Dedekind. Thanks to the efforts of the latter (and to H. Weber), Riemann's unpublished works were found after his death and issued in a volume of his collected works, revealing the splendor of his talent to mathematicians. Riemann's work on Abelian functions developed the theme of his doctoral dissertation. His highly original and fruitful idea was that multivalued functions (for example, $\sqrt{z}$) could be represented as single-valued functions, not over the complex plane but rather over a special (Riemann) surface defined by the singularities of the function.

To give the reader an appreciation of the beauty of this construction we shall begin by constructing the Riemann surface for the two simplest cases

$$w = \sqrt{z} \tag{4.1}$$

$$w = \sqrt{(z-a_1)(z-a_2)(z-a_3)}. \tag{4.2}$$

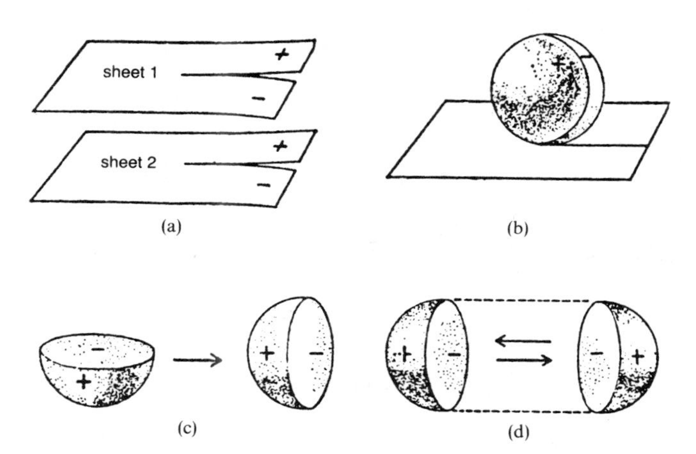

Figure 4.1: A simply connected Riemann surface.

1) The Riemann surface of the function $w = \sqrt{z}$. We choose some point $z_0 = r_0 e^{i\phi_0}$; then $w = \sqrt{r_0}e^{i\phi_0/2}$. Given a circuit of the origin of the coordinates along a closed contour ($\phi = \phi_0 + 2\pi$), the function w changes sign: $w = -\sqrt{r_0}e^{i\phi_0/2}$. Thus, for one and the same point $z_0 = r_0 e^{i\phi_0}$ (to which we return after a circuit of the origin) we obtain two different values of w—the function is two-valued. To get around this difficulty, we cut the complex z-plane along the positive part of the real axis. As long as $w(z)$ traverses paths that do not intersect this cut, we have two single-valued branches of $w(z)$:

$$w(z) = \sqrt{r}e^{i\phi/2}, \quad \text{when } 0 \le \phi < 2\pi;$$
$$w(z) = -\sqrt{r}e^{\phi/2} \quad \text{when } 2\pi \le \phi < 4\pi.$$

Now we take two copies of the z-plane with cuts (Fig. 4.1). Each cut has two edges: an upper and a lower one (imagine that the cut has a certain width). We indicate the edge of the upper quadrant with a positive sign and the lower with a negative. In order to continue a branch of $w(z)$ across the cut, it is necessary that its sign on the different edges coincide. Now we glue the two sheets together: the upper edge of the first sheet to the lower edge of the second and, correspondingly, the lower edge of the first sheet to the upper edge of the second. Given a passage across the cut, we go from one sheet to the other. The following remarkable situation results: a moving point that begins at point z_0 on the first sheet and makes a circuit about the origin along a closed contour passes over to the second sheet, and at point z_0 the function w will equal $-\sqrt{z_0}$. Thus, on each sheet, we obtain a single value of the function w. Riemann's basic idea is as follows: one can consider a multivalued function on a complex plane as single-valued on the corresponding multisheeted Riemann surface. Unfortunately, such gluing without intersection cannot be done in three dimensional space, but one can prove that such a Riemann surface is topologically equivalent to a sphere. By using stereographic projection one can map each sheet onto a sphere (the Riemann

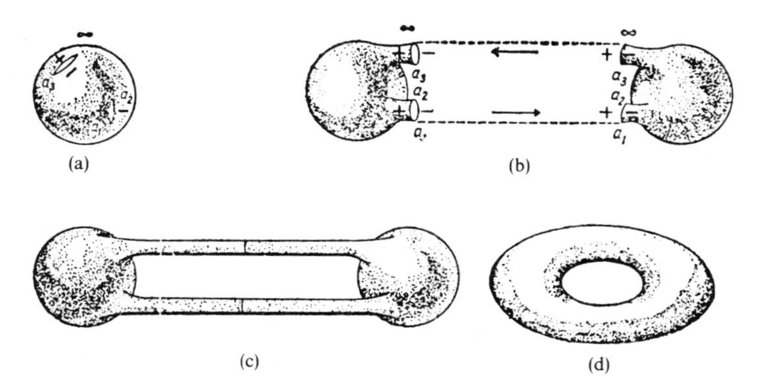

Figure 4.2: A doubly connected Riemann surface.

sphere). A cut on the plane becomes a cut along the meridian (Fig. 4.1b). We move the two edges of the cut so that we obtain a hemisphere (Fig. 4.1c). Having carried out a similar operation with the second sheet and having glued together both hemispheres, observing the law of signs, we obtain a new sphere (Fig. 4.1d). The same result can be reached by gluing together the corresponding edges of two sheets. Because these operations are topologically equivalent, the Riemann surface of the function $w(z)$ is homeomorphic to the sphere.[1]

2) Construction of the Riemann surface of more complex functions, for example, $w = \sqrt{(z-a_1)(z-a_2)(z-a_3)}$. On the sphere we make cuts from a_1 to a_2 and an additional cut from a_3 to a_4, where the point a_4 is adjoined at infinity (Fig. 4.2a); we separate these cuts, stretch them out (Fig. 4.2b) and, observing the law of signs, glue the two pieces together (Fig. 4.2c). This surface is topologically equivalent to a torus (Fig. 4.2d) or to a sphere with a handle, which is the same thing. In general the Riemann surface of the function $w = \sqrt{(z-a_1)\cdots(z-a_{2g})}$ is a sphere with $g-1$ "handles." If the number of points a_i is odd, we adjoin another point $a_{i+1} = \infty$. The number of handles represents an important topological invariant: the *genus* of the surface. Two surfaces are topologically inequivalent if they have different genera. Riemann explicitly recognized the topological nature of his constructions. His work and the subsequent work of Poincaré, became the foundation of topology.

It should be noted that Riemann confined himself to representing a Riemann surface as a collection of sheets with a certain rule gluing them together. The fact that the Riemann surface of an algebraic function (that is, a function of the type $f(z_1, z_2) = P_n(z_1)z_2^n + \cdots + P_0(z_1)$, where the coefficients $P_k(z_1), \ldots$ are polynomials in z_1) is topologically equivalent to an arbitrary orientable closed two-dimensional surface was clarified later, basically thanks to Klein's 1881/1882 paper "Über Riemanns Theorie der algebraischen Funktionen und ihrer Integrale" (On Riemann's the-

[1] In topology a one-to-one continuous mapping having a continuous inverse is called a *homeomorphism*.

ory of algebraic functions and their integrals).

Riemann obtained an important formula linking the number of sheets n, the number of branch points b (with corresponding multiplicities), and the genus of the Riemann surface of an algebraic function:

$$\tilde{g} = \frac{b}{2} - n + 1. \tag{4.3}$$

This formula shows that the genus of the hyperelliptic surface

$$w = \sqrt{(z - a_1) \cdots (z - a_{2g})}$$

equals $g - 1$ (the number of sheets is $n = 2$; the branch points $a_1, \ldots, a_{2g}$ all have multiplicity 1). One can construct a one-to-one continuous mapping of any closed orientable two-dimensional surface onto a sphere with a certain number of "handles." From this it follows that any closed orientable surface is the Riemann surface of an algebraic function.

We emphasize that Riemann surfaces are orientable. This means that a Riemann surface cannot be a Möbius strip, a projective plane, a Klein bottle, or any other nonorientable surface. The orientability of a Riemann surface follows from the fact that a complex structure exists on it: a Riemann surface can be represented as a complex curve in two-dimensional complex space $\{(w, z) \in \mathbb{C}^2 : w = f(z)\}$. The proof of this fact is not complicated but requires the introduction of new concepts, and we shall not give the details at this point.

The introduction of Riemann surfaces greatly advances the study of algebraic functions and their integrals—the theory of Abelian integrals. This theory is the subject of study in the classic works of Abel and Jacobi. It owes its origin to the work of Jakob and Johann Bernoulli (1657–1705 and 1667–1748 respectively), Euler, and Legendre; it is one of the classical areas of analysis. Integrals that could not be expressed in elementary functions arose in the seventeenth century in solving the problem of calculating the arc length of an ellipse. For that reason such integrals came to be called *elliptic*. The problem of inverting elliptic integrals was solved by Abel and Jacobi. This problem is stated as follows: *Given an integral of the form*

$$z = \int_0^u \frac{du}{\sqrt{P_k(u)}}, \tag{4.4}$$

where $P_k(u)$ is a polynomial of the third or fourth degree, find the function u in terms of z. Elliptic integrals are associated with the torus—a Riemann surface of genus 1. It turns out that, in this case, one can represent the function u as a single-valued doubly periodic function of z.

Abel, however, had considered much more general algebraic integrals than elliptic. He showed that a finite sum of m integrals of the type

$$\int_{x_0, y_0}^{x_1, y_1} R(x, y)\, dx + \cdots + \int_{x_0, y_0}^{x_m, y_m} R(x, y)\, dx,$$

NIELS HENRIK ABEL

(where $R(x, y)$ is a rational function and x and y satisfy a relationship $f(x, y) = 0$, where $f(x, y)$ is a polynomial) can be expressed by only g such integrals and a certain number of rational and logarithmic terms. Here g, the genus of the algebraic function, does not depend on m (for hyperelliptic integrals $f(x, y) = y^2 - P_k(x)$, and the genus of the algebraic function is the already familiar genus of the Riemann surface). If the integral involves the square root of a polynomial of fifth degree or higher (a hyperelliptic integral), so that the genus will be larger than 1, there will be two or more upper limits of integration in the sum of the integrals, and hence one cannot expect any simple inversion such as occurs for elliptic integrals. Abel himself did not pose the inversion problem in the general case. Jacobi had the idea of supplementing a single hyperelliptic integral by others of the same kind so as to obtain a number of equations equal to the number of upper limits. The resulting problem, posed in 1831, was known as the *Jacobi inversion problem*. Although Jacobi had some success in solving the problem of inversion in the case of hyperelliptic integrals, it was clear that new ideas were needed for a complete solution.

Abel's theorem makes it possible to reduce the study of integrals of arbitrary algebraic functions—Abelian integrals—to several special cases (depending on the type of singularities), for example, we can consider integrals for which logarithmic terms are absent, and the like. We see that in general the inversion requires the study of multivalued functions on an arbitrary Riemann surface.

The honor of solving this problem belongs to Riemann. He began his investigation with the construction of a general theory of multivalued analytic functions on a Riemann surface. The basic method used by Riemann was already contained in his doctoral dissertation. He showed first that it is possible, using a system of cuts, to convert a multiply connected domain into a simply connected one and, second, having determined the behavior of a function during a passage across the cuts (jumps), to reduce the problem to the study of single-valued functions with a given type of singularity.

Riemann succeeded in showing that the existence of multivalued functions with a given type of singularities depends on the topology of the Riemann surfaces. Consider, for example, the following Abelian integral:

$$w = \int_0^z R(u, z)\, dz, \tag{4.5}$$

where u and z are related by a polynomial equation $f(u, z) = 0$. The integrand is called an Abelian differential and the function $R(u, z)$ is an Abelian function.

One can pose the following question: *Do there exist nonconstant Abelian functions having no poles on a given Riemann surface?* (The integrals corresponding to such functions are called integrals of the first kind). That such functions must be multivalued follows from a theorem of J. Liouville (1809–1882), stating that a single-valued analytic function without singularities (poles) on a closed Riemann surface is constant. Riemann obtained the following result: *On a surface of genus g, there exist g linearly independent integrals of the first kind.* (It follows from this, for example, that on a sphere there are no regular Abelian functions.)

Riemann solved an analogous problem for Abelian functions that approach infinity at a finite number of points. Here the problem is significantly more complicated. For example, consider the problem of the existence of Abelian differentials of the second kind, that is, having a representation

$$\left(\frac{a_2}{z^2} + \cdots + \frac{a_k}{z^k} + \phi(z)\right) dz = f(z)\, dz$$

at every point (where $\phi(z)$ is a holomorphic function). It is impossible to prescribe completely arbitrary poles for a a meromorphic function $f(z)$ whose differential is an Abelian differential of the second kind. It follows from Liouville's theorem, that the function $f(z)$ must have at least one pole.

Riemann succeeded in obtaining a most important result, now known as Riemann's inequality: *The number r of linearly independent meromorphic functions with poles of order not greater than n_k at m distinct points P_k, $(k = 1, \ldots, m)$ is not less than $\sum n_k - g + 1$, where g is the genus of the surface.*

In 1864 Gustav Roch (1839–1866), a student of Riemann who also died at an early age, succeeded in strengthening this result. It turns out that

$$r = \sum n_k - g + 1 + i[a],$$

where $i[a]$ is the number of linearly independent differentials with zeros at the points P_k of order at least n_k and having no poles on the Riemann surface. This is the famous Riemann–Roch theorem. At the present time numerous multidimensional generalizations of the Riemann–Roch theorem play an important role in various branches of algebraic geometry, analysis, and topology.

Another discovery that Riemann made while studying Abelian functions is connected with the fine structure of Riemann surfaces. We have repeatedly stressed the topological background of Riemann's research, but it turns out that there exists a whole class of transformations that preserve the genus of a surface and, consequently its topology, but which lead to finer distinctions. We are talking about birational (Cremona) transformations.

Consider the surface

$$f(w, z) = 0. \tag{4.6}$$

If $w_1 = R_1(w, z)$ and $z_1 = R_2(w, z)$ are rational functions and the inverse transformation also is given by rational functions, then the function $f(w, z)$ leads to a new function $F(w_1, z_1)$. The corresponding surfaces $f(w, z) = 0$ and $F(w_1, z_1) = 0$ are said to be *birationally equivalent*. A necessary condition for birational equivalence of two surfaces is that they have the same genus, but this condition is far from sufficient.

It turns out that among Riemann surfaces of genus $g > 1$ there exists a $(3g - 3)$-dimensional family of birationally inequivalent Riemann surfaces. For $g = 1$, the family of birationally inequivalent Riemann surfaces has dimension 1, but for $g = 0$ (a sphere), there are no birational invariants. It is interesting to note that this number corresponds to the number of conformally inequivalent Riemann surfaces. If we turn again to algebraic equations, we obtain the following result: *Algebraic equations of the type of Eq. (4.6) having genus g depend on* $3g - 3$ *complex parameters*. These parameters are called the *moduli* of the algebraic curve (or the Riemann surface). The number $3g - 3$ defines the dimension of the space of moduli of the space of conformally inequivalent Riemann surfaces. A more detailed description of this space—for example, the introduction of a metric in it—is a very complex problem. It was partially solved only in the early 1940s. In the concluding part of the article, Riemann gave a general solution to the problem of inversion of Abelian integrals defined on a surface of genus g.

To this end he perfected the theory of multivariable θ-functions on Riemann surfaces, which generalize Jacobi's elliptic θ-functions. (These functions had been introduced in 1847 by Adolf Göpel (1812–1847) and Johann Georg Rosenhain (1816–1887).) The dimension of the space of θ-functions is determined by the genus of the Riemann surface. The basic concepts of the theory of θ-functions introduced by Riemann have remained unchanged down to the present and constitute the foundation of the whole modern theory of θ-functions, which are a component of the theory of complex and algebraic varieties. It suffices to mention the condition for convergence of θ-series, that the real part of the period matrix of the Abelian differentials (the "Riemann matrix") be negative-definite, or the condition for a complex torus to be algebraic—the Riemann–Frobenius condition.

But, in a manner typical of all of Riemann's work, the introduction of general structures was not an end in itself for him. He immediately applied the apparatus of θ-functions to solve the Jacobi inversion problem, reducing it to the problem of determining the zeros of the θ-functions. The answer depends on conditions under which a θ-function is not identically zero on the Riemann surface. In the "Theorie der Abel'schen Functionen" and its sequel "Über das Verschwinden der θ-Functionen" (1865) Riemann obtained a necessary and sufficient condition for the vanishing of θ-functions that gave a complete solution of the problem of inverting Abelian integrals.

In proving the fundamental theorem on the vanishing of θ-functions Riemann derived a number of important identities for θ-functions—the so-called *Riemann θ-relations*—as a secondary result. These relations play a vital role in modern research in the theory of Riemann surfaces, in particular in solving the Schottky problem. This famous problem, which originated with Riemann, was clearly formulated by the German mathematician F. Schottky (1851–1935) in 1903 and consists of the following: *Suppose given matrices of a Riemann surface of genus g, which form a family of complex matrices depending on $3g - 3$ parameters, $g > 1$ (the moduli of the surface), while the complete family of Riemann matrices, which satisfy the Riemann–Frobenius condition, form a family of complex symmetric matrices B_{ij} of dimension $\frac{g(g+1)}{2}$. What conditions must be imposed on the Riemann matrices B_{ij} so that they are the period matrices of Abelian differentials of a certain Riemann surface of genus g?* The answer is quite simple when $g \leq 3$; in this case any matrix B is the period matrix of a corresponding Riemann surface. But when $g \geq 4$, the answer is highly nontrivial (the dimension of the space of matrices B increases considerably faster than the dimension of the space of moduli). A full and effective solution of Schottky's problem was obtained comparatively recently by T. Shiota (1986) and involves remarkable achievements in the theory of nonlinear evolution equations. For more details on these equations see the chapter "Soliton Particles" below.

The fate of this remarkable work is highly surprising. Weierstrass, Riemann's chief rival in developing the theory of Abelian functions, was deeply disturbed by these results. As Klein reported:

> ...when Weierstrass submitted his first treatise on general Abelian functions to the Berlin Academy in 1857, Riemann's paper on the same theme appeared in *Crelle*, Volume 54. It contained so many new and unexpected ideas that Weierstrass withdrew his paper and in fact published no more.[2]

Considering that Weierstrass had his own solution to the Jacobi inversion problem for general hyperelliptic integrals and had obtained a number of highly interesting relationships for general elliptic functions, his scholarly fastidiousness is completely

[2] Klein, F. *Development of Mathematics*, p. 264.

KARL WEIERSTRASS

unprecedented. For a number of years, this work of Riemann was considered his most significant contribution to mathematics. But Weierstrass soon pointed out a serious gap in it, which placed all of Riemann's fundamental results in jeopardy.

The fact is that, already in his doctoral dissertation, and especially in the paper "Theory of Abelian functions," in order to assert the existence of a required function with a given type of singularities, Riemann used the variational principle we have already discussed—the Dirichlet principle, which assumes the solution of a variational minimization problem. Weierstrass expressed doubts about this principle to Riemann, and (after Riemann's death) he showed that there were similar problems in the calculus of variations that have no solution. Thus the particular result required by Riemann would need a special proof, which Riemann had not given.

Klein describes the reaction to Weierstrass criticism as follows:

> The majority of mathematicians turned away from Riemann... Riemann had been of a quite different opinion. He fully recognized the justice and correctness of Weierstrass's critique; but he said, as Weierstrass once told me, "that he appealed to the Dirichlet principle only as a convenient tool that was close at hand, and that his existence theorems are still correct."[3]

[3]*Ibid.*, pp. 247–248.

Weierstrass himself was also convinced of this last assertion. He encouraged his student Hermann Amandus Schwarz (1843–1921) to make a thorough study of Riemann's existence theorem and to seek other proofs, which Schwarz succeeded in doing.

In 1869, immediately after Weierstrass's critical remarks appeared in print, Schwarz proved the existence of a solution of the Dirichlet problem without using the variational method. His alternative method consisted of the following: he first solved the Dirichlet problem for the disk, using the Poisson integral to construct a harmonic function $u(x, y)$ in the disk, taking prescribed values on the boundary. Then he showed how to pass to an arbitrary domain obtained as the union of a finite number of disks. Another remarkably interesting proof, proposed by Poincaré, was based on potential theory—the method of *balayage* (sweeping out).

Riemann's "mistake" had yet another remarkably useful consequence. It stimulated specialists in algebraic geometry to find a purely algebraic proof of Riemann's theorem, specifically the Riemann–Roch theorem. The outstanding work of Rudolph Friedrich Alfred Clebsch (1833–1872), who invented the term "genus of a surface," is in this area. Paul Gordan (1837–1912), Max Noether (1844–1921), and finally (in 1899) Hilbert succeeded in giving a proof of the variational principle. It is difficult to recall another example in the history of nineteenth-century mathematics when the search for a rigorous proof led to such productive results.

Physicists were completely convinced by Riemann's work. As evidence of this fact we cite an excerpt from a paper of A. Sommerfeld (1868–1951), "Klein, Riemann and Mathematical Physics:"

> Riemann's dissertation was at first strange to his mathematician contemporaries, who reviewed it as if it were a book published for the family. The fact that it was closer in its way of reasoning to physics than to mathematics is attested to by a story of one of my colleagues. Once he spent his vacation together with Helmholtz and Weierstrass. Weierstrass had taken Riemann's dissertation along on holiday in order to deal with what he felt was a complex work in quiet circumstances. Helmholtz did not understand what complications mathematical specialists could find in Riemann's work; for him Riemann's exposition was exceptionally clear.

Why was Riemann's explanation so clear to physicists when it presented such difficulties to mathematicians? The reason is certainly not the obtuseness of mathematicians and the brilliance of physicists, but rather the different standards of proof demanded by the two professions.

There exists a remarkable interpretation of the theory of functions on Riemann surfaces, which owes its origin to Helmholtz and which explains why physicists had such confidence in the validity of Riemann's results. One can interpret the theory of analytic functions on Riemann surfaces as a problem of physics. We shall show, in fact, that the theory of a stationary two-dimensional ideal incompressible fluid on a surface leads to the theory of analytic functions.

RUDOLPH FRIEDRICH ALFRED CLEBSCH

Consider a stationary flow of fluid U on the (x, y) plane. The velocity of the flow at each point has an x-component $P(x, y)$ and a y-component $Q(x, y)$. Through a small rectangle with sides Δx, Δy, the following mass of fluid flows in unit time (the density of the fluid is constant and equal to 1):

$$\int_0^{\Delta y} \{P(x+\Delta x, y+h) - P(x, h+h)\}\, dh + \int_0^{\Delta x} \{Q(x+l, y+\Delta y) - Q(x+l, y)\}\, dl. \tag{4.7}$$

Approximating an arbitrary domain Ω by rectangles and applying Green's theorem, we deduce that the integral (4.7) is equal to

$$\iint \left(\frac{\partial P}{\partial x} + \frac{\partial Q}{\partial y}\right) dx\, dy. \tag{4.8}$$

Since the fluid is incompressible and there are no sources or sinks in the domain Ω, it follows that expression (4.8) is equal to zero. A stronger assertion is also valid: the divergence of the flow U equals zero:

$$\operatorname{div} U = \frac{\partial P}{\partial x} + \frac{\partial Q}{\partial y} = 0. \tag{4.9}$$

The *circulation* of the flow along a curve C is defined to be the integral $\int P\, dx + Q\, dy$. If this integral along any closed curve is equal to zero, the flow is called *irrotational.* For any simply connected domain this implies that the expression $P\, dx + Q\, dy$ is the total differential of a function $u(x, y)$, which in turn is harmonic.

The function $u(x, y)$ is called the *velocity potential* of the flow, a concept introduced by Helmholtz. The curves $u(x, y) = \text{const}$ are called the *equipotential lines*, and the tangent to an equipotential line forms an angle α with the x-axis such that $\tan\alpha = -(\partial u/\partial x)/(\partial u/\partial y)$ if $\nabla u \neq 0$. The velocity vector of the flow forms an angle β with the x-axis, where $\tan\beta = (\partial u/\partial y)/(\partial u/\partial x)$, that is, the direction of the flow is orthogonal to the equipotential lines and points in the direction of most rapid decrease of the function u.

Recall that a harmonic function $u(x, y)$ determines an analytic function $f(x, y) = u + iv$, where v, the *harmonic conjugate* of the function u, is defined by the Cauchy–Riemann equations (see formula (2.1)). The function $f(z)$ is called the *complex potential* of the flow. The tangent to a curve $v = \text{const}$ forms an angle γ with the x-axis and

$$\tan\gamma = -\frac{\frac{\partial v}{\partial x}}{\frac{\partial v}{\partial y}} = \frac{\frac{\partial u}{\partial y}}{\frac{\partial u}{\partial x}} = \tan\beta,$$

that is, the current u flows along the curve $v = \text{const}$. These curves are called *streamlines*. The condition that $(\partial u/\partial x)^2 + (\partial u/\partial y)^2 = 0$ is equivalent to the condition that $f'(z) = 0$. This implies that the streamlines are orthogonal to the equipotential lines, except when $f'(z) = 0$.

This physical analogy provides a completely intuitive interpretation of the properties of analytic functions. For example, if an analytic function $f(z)$ has $f'(z_0) = 0$ at a point z_0, then the curves $u = \text{const}$ and $v = \text{const}$ no longer intersect at a right angle at the point $z_0 = x_0 + iy_0$. Such points are called *stationary points*. For example, for the function

$$f(z) = a_0 + a_2 z^2$$

the curves $u = \text{const}$ and $v = \text{const}$ intersect at an angle of $\frac{\pi}{4}$.

With equal success we can study arbitrary singularities of analytic functions. Let us consider a flow with potential $f(z)$ whose derivative $f'(z)$ is a rational function, that is, its only singularities are poles (for example, $(z - a_0)^{-k}$). Then the function itself can be represented in a neighborhood of a singularity as

$$f(z) = A\log(z - z_0) + A_1(z - z_0)^{-1} + \cdots + \phi(z), \tag{4.10}$$

where $f(z)$ is a function without singularities.

The singularities of the flow defined by the function $f(z)$ can be constructed from the singularities of the flows defined by the individual terms of (4.10). Let us consider the behavior of the logarithmic term. We first suppose that A is a real number. We choose a disk of radius r about the point $z_0 : z = z_0 + re^{i\phi}$ and set

$$A\log(z - z_0) = u + iv.$$

Separating the real and imaginary parts, we obtain $u = A\log r$, $v = A\phi$. The streamlines $v = \text{const}$ are rays emanating from the point z_0, while the equipotential lines $u = \text{const}$ are circles with center z_0 (see Fig. 4.3). Thus, the point z_0 is either a source (Fig. 4.3a) or a sink (Fig. 4.3b) for the flow, depending on the sign of A (the liquid either flows out of or into the point z_0). If A is a purely imaginary number, we obtain the conjugate flow: $A = iB$, $u = -B\phi$, $v = B\log r$ for which the stream lines are circles. Such flows are called *vortices*. The direction of the motion (clockwise or counterclockwise) is determined by the sign of B.

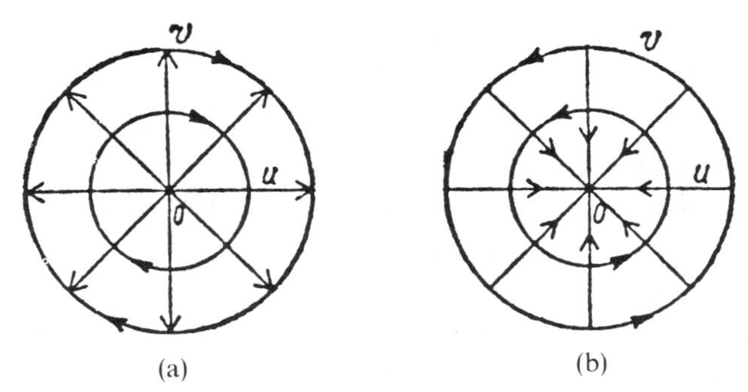

Figure 4.3: A source (a) and a sink (b).

We have thus obtained a remarkable result: *The singularities of an analytic function $f(z)$ can be described in terms of the flow of a fluid with a certain number of sources, sinks, vortices, etc.* Sommerfeld called this whole circle of ideas *physical mathematics*. As he said, "Here it is not mathematics serving the interests and problems of physics, but rather physics inspiring and governing mathematical ideas."

At the end of his work on the theory of Abelian functions, in connection with a definition of Abelian integrals in terms of special functions (θ-functions), Riemann wrote, "In order to carry out all of the reasoning necessary for this purpose, one must obviously rely on the further development of the theory of functions that satisfy linear differential equations with algebraic coefficients; I propose to take this up in the near future... ."[4] To this subject he devoted two papers ("Two general theorems on linear differential equations with algebraic coefficients" and "On integrals of a second-order linear differential equation in a neighborhood of a branch point"), which he wrote in 1857–1858, and a course of lectures on hypergeometric series. All of this was published after his death.

In the first of these papers Riemann formulated the following problem from the theory of linear equations. Consider a system of homogeneous linear differential equations in the complex plane:

$$\frac{dy_i}{dz} = \sum_{j=1}^{n} A_{ij}(z)y_j, \tag{4.11}$$

where the functions A_{ij} are rational in z. Solutions of the system (4.11) turn out to be multivalued functions. The singularities of each solution of Eq. (4.11) are determined by the poles of the matrix $A(z)$: $a_1, \ldots, a_k$, $a_0 = \infty$. The solution naturally changes after a circuit about a singularity. Riemann found a condition guaranteeing that the solution following a circuit of the singular point differs from the original one only by a constant matrix. If after a circuit about the point a_1 the solution $\eta_1 = (y_{11}(z), \ldots, y_{n1}(z))$ changes to the vector $\sum b_{1j}^{(1)}\eta_j$, while the solution η_i changes to $\sum b_{ij}^{(m)}\eta_j$ after a circuit about the point a_m, then the matrices

[4]From Riemann's "Theorie der Abel'schen Functionen."

$B^{(0)}, \ldots, B^{(m)}$ are nonsingular and satisfy Riemann's relation:

$$B^{(0)} \times B^{(1)} \times \cdots \times B^{(m)} = E, \tag{4.12}$$

where E is the identity matrix.

In modern terminology relations of type (4.12) define a monodromy mapping, that is, a mapping

$$\pi_1(\widetilde{C} \setminus \{a_0, \ldots, a_m\}) \to GL(n, \mathbb{C}), \tag{4.13}$$

where the order n of the group GL$(n, \mathbb{C})$ is defined by the dimension of the fundamental matrix of solutions of the system (4.11). We use the standard mathematical notation: GL$(n, \mathbb{C})$ is the set of nonsingular complex $n \times n$ matrices, $\widetilde{\mathbb{C}}$ is the extended complex plane, completed by adjoining the point $a_0 = \infty$ (the "Riemann sphere"), and $\pi_1(X)$ is the fundamental group (the first homotopy group) of the set X. The matrices $B^{(i)}$ are called *monodromy matrices* and are generated by circuits about the points a_i over simple loops (that is, closed paths containing no other singularities).

In this same work, Riemann posed the converse problem: *Given a system of points $a_0, \ldots, a_m$, does there always exist a system of equations of type (4.11) having given singularities and given transformation matrices satisfying (4.12)?* Riemann made a conjecture about the form of such equations, but did not produce a general proof. In his lectures on hypergeometric series, he considered the case, $n = m = 2$.

Solutions of equations of type (4.11) constitute a very large class of functions, for example, the majority of the special functions of mathematical physics, in particular Bessel functions, hypergeometric functions, etc. Riemann's work remained unpublished for almost twenty years. Not knowing of Riemann's work, the well-known mathematician Lazarus Fuchs (1833–1902), a student of Weierstrass, set to work on this series of questions in 1865. He gave a detailed classification of the singular points of equations of type (4.11). The most important class of such equations with matrix $A_{ij}(z)$ having simple poles as singularities came to be called *Fuchsian equations*, a term casually coined by Poincaré.

Riemann's basic problem on the existence of equations with given monodromy matrices and singularities remained unsolved. It was considered so difficult that Hilbert included it as No. 21 in his famous list of "Mathematical Problems."

In a speech delivered at the Second Mathematical Congress in Paris in 1900, Hilbert posed twenty-three problems, the solution of which he considered important for the development of mathematics. Of these problems, three were associated with Riemann's name. Problem No. 21, just mentioned, was one of them, as were No. 20 (the proof of Dirichlet's variational principle, solved by Hilbert himself) and No. 8 (associated with the "Riemann Hypothesis," which we will discuss later).

Concerning Hilbert's Problem No. 21, the history of its solution is very engrossing. Until recently it was believed that it had been solved in 1908 by the Serbian mathematician J. Plemelj (1873–1967). Since no doubts had been raised about Plemelj's proof, research was directed mainly toward finding effective methods of constructing equations from a given monodromy group. In particular a detailed analysis of branch points had been carried out by the Russian mathematician Ivan Aleksandrovich

LAZARUS FUCHS

Lappo-Danilevskii (1896–1931), who developed the machinery of analytic functions of matrices specifically for the purpose. The Riemann problem was extended to an arbitrary Riemann surface by the German mathematician H. Röhrl, in 1957. But the most dramatic events occurred in 1989, when the Moscow mathematician A. Bolibruch constructed a counterexample to Plemelj's theorem. It turned out that given any $m > 3$, any set of points $a_1, \ldots, a_m$ and any $n \geq 3$, there exists a representation (4.13) not realized by any Fuchsian system. This remarkable result forced a re-examination of this entire area of differential equations. Additional interest in systems of Fuchsian type arose as the result of a number of novel applications in modern theoretical physics, in particular, in two-dimensional conformal field theory, where correlation functions, the main object of study, satisfy a special system of differential equations of Fuchsian type.

Interesting problems associated with equations of Fuchsian type arise in contemporary theoretical physics. The basic object in describing the interaction of elementary particles—the scattering amplitude—is an analytic function depending on the momentum of interacting particles and having singularities of quite complicated structure which are determined by the basic physical requirements. In complex space scattering amplitudes have not only poles but also branch points and even singularities

IVAN ALEKSANDROVICH LAPPO-DANILEVSKII

along entire curves—Landau singularities. The basic method of investigating scattering amplitudes is a representation in the form of a series in perturbation theory. The computations are carried out by the method of Feynman diagrams.

Generalizations of Riemann's methods play an important role in the problem of classifying singularities, since Feynman integrals and other important functions (for example, Dirac's well-known δ-function) are generalized solutions of equations of Fuchsian type with analytic coefficients.

The range of Riemann's scholarly interests was enormous. At the same period when he was carrying on these investigations, as his friend and biographer Dedekind showed,

> His thoughts again turned to natural philosophy, and one evening after a strenuous hike, he reached for Brewster's book *The Life of Newton*,[5] and for a long time he talked admiringly about Newton's letter to Richard Bentley (1662–1742) in which Newton himself affirmed the impossibility of direct action at a distance.[6]

[5]Brewster, D. *Memoirs of the Life, Writings, and Discoveries of Sir Isaac Newton*, Edinburgh: Constable, 1855.

[6]Riemann, B. *Gesammelte Mathematische Werke*, p. 553.

It is possible that this very remark by Newton suggested to Riemann the idea of introducing the concept of action at a distance in electrodynamics. In the paper "On the subject of electrodynamics," presented by Riemann on February 10, 1858 to the Royal Göttingen Scientific Society, he formulated his result in the following words:

> ...I allow myself to make a remark which introduces a close connection between the theory of electricity and magnetism and the theory of light and radiant heat. I have established that the electrodynamic action of galvanic flows can be explained by assuming that the action of electric mass on other bodies does not occur instantaneously but propagates in its direction with constant speed (the speed of light within the limits of possible observational error). Given this assumption, the differential equation for the propagation of electric force is the same as the equation for the propagation of light and radiant heat.[7]

Riemann's result is striking when we reflect that this announcement was made more than seven years before the discovery of the famous Maxwell equations and in the city where Weber's law held absolute sway. According to this law, between moving particles with charges e and e', there is an instantaneous force:

$$F = ee'\left\{\frac{1}{r^2} - \frac{1}{(c')^2 r^2}\left(\frac{dr}{dt}\right)^2 + \frac{1}{(c')^2 r}\left(\frac{d^2 r}{dt^2}\right)\right\}$$

(c' is the Weber–Kohlrausch constant, equal to $c\sqrt{2}$, where c is the speed of light). Unfortunately, in his calculations there were several technical mistakes, as his audience pointed out, especially Kohlrausch. Riemann evidently accepted the opinion of such a well-known experimenter and withdrew his paper. Like many other works by Riemann, it was published in his collected works after his death.

Weber's law was considered the last word in science in Germany for a long time, until the remarkable experiments of Heinrich Hertz (1857–1894) confirmed Maxwell's theory.

Despite his brilliant achievements, scientific recognition of Riemann proceeded very slowly. For that reason the support of Dirichlet and his younger colleague Dedekind provided great satisfaction to Riemann and strengthened his confidence in himself.

Dirichlet's name is so often recalled along with that of Riemann that it is necessary to give at least a brief portrait of this multifaceted mathematician—after Gauss, Germany's leading mathematician of the first half of the nineteenth century.

Peter Gustav Lejeune-Dirichlet, of French ancestry, was born in Germany. From the age of seventeen he lived in Paris where he became acquainted with Fourier. His first significant work, which we have already mentioned, was devoted to Fourier series. In 1827 the eminent natural scientist Alexander von Humboldt (1769–1859) invited Dirichlet to Prussia. Von Humboldt, like his brother Wilhelm, possessed a

[7] *Ibid.*, p. 288.

PETER GUSTAV LEJEUNE-DIRICHLET

rare combination of scholarly and administrative talents. Both brothers had an exceptionally beneficial influence on the development of science in Germany. Alexander von Humboldt, being a personal friend of King Friedrich Wilhelm III, had great prestige in governmental circles and used his influence in the interests of science. The invitation to Dirichlet was by no means his only good deed.

After two years as assistant professor at Breslau (now Wrocław), Dirichlet moved to Berlin where he held posts consecutively as assistant professor, then associate professor, and finally, from 1839 on, as full professor.

Dirichlet obtained first-class results in the most varied branches of mathematics. Aside from the theorems already mentioned in the theory of functions and the Dirichlet principle, which played an outstanding heuristic role in the development of analysis, Dirichlet made a major contribution to number theory.

Here we recall only his main achievement—the application of analytic methods for solving arithmetic problems. In the course of proving that a general arithmetic sequence contains an infinite set of prime numbers he introduced a series of the type

$$\eta(s) = \sum \frac{a_n}{n^s}.$$

Series of this form are called *Dirichlet series*, and a generalization of them plays a most important role in the contemporary theory of numbers. One of Dirichlet's theorems involves the distribution of prime numbers in arithmetic progressions. Despite its highly nontrivial proof, the formulation of Dirichlet's theorem is elementary: *If a and m are relatively prime positive integers, there exists an infinite set of prime numbers p, such that* $p \equiv a \pmod{m}$.

Dirichlet's place in science is determined not only by his own scholarly attainments: his methods of teaching had a remarkably significant influence on German science and gradually on the whole world. Klein gave an excellent characterization of Dirichlet's style:

> He knew how to present his clear inner perceptions so convincingly in words alone that they seemed to proceed from their premises in a self-evident way... For Dirichlet, teaching and research were inseparably bound together... .[8]

I will cite Klein once more in order to give an idea of Dirichlet as a person:

> His single goal, which he strove for with his whole being, was clear insight into the ideal coherence of mathematical thought, a goal that led him to renounce worldly influence and success. As is so often the fate of quiet men who seek and find satisfaction within themselves, it was his destiny to be surrounded by aggressive, strongly outwardly directed men. Dirichlet married into the rich and gifted Mendelssohn family: his wife was Rebekka, a sister of Felix Mendelssohn. Since this family was one of the most brilliant centers of social life in Berlin at that time, Frau Dirichlet was able to gather about her, in the brief Göttingen period, all the people most interested in science and art, creating a lively and cultivated social life. It is said that Dirichlet took part in the social arrangements at his house only in a very reticent and retiring way. The incessant choppy sea of dazzling intellects around him could not in the least have corresponded to the deeper sea-swell of his own spirit.[9]

Dirichlet played an exceptional role in Riemann's life. In Dedekind's words, "From the very beginning he felt the liveliest personal attraction toward Riemann." In temperament as well as in his scientific interests, Dirichlet was remarkably close to Riemann. For Riemann in these years scientific support was not the only concern. He was forced to lead a very difficult life. Family misfortunes befell him: in 1857 a brother died, and soon after, a younger sister. These events, in addition to a serious nervous exhaustion brought on by intensive scholarly work, brought about depression. Only the help of Dirichlet, Dedekind, and his loving sisters, as well as a short trip to the mountains, brought Riemann back to a normal life.

In the fall of 1858 the Italian mathematicians Enrico Betti (1823–1892), Felice Casorati (1835–1890), and Francesco Brioschi (1824–1897) came to Göttingen, and

---

[8]Klein, F. *Development of Mathematics*, pp. 87–88.

[9]*Ibid.*, p. 90.

Riemann became friends with them. They shared their results; in particular, Riemann communicated to Betti several facts from topology (the concept of connectivity for n-dimensional manifolds). Their interest in his work gave Riemann great satisfaction. The end of 1858 marked a long-awaited breakthrough in Riemann's life. His works at last became well-known and acknowledged.

Chapter 5

Full Professor in Göttingen

"Nothing stimulates great minds to work on enriching knowledge with such force as the posing of difficult but simultaneously interesting problems."

JOHANN BERNOULLI

ON May 5, 1859 Dirichlet died after a serious illness. The government no longer wavered in its choice of a successor. On July 30 of that year Riemann was made a full professor of Göttingen University. From that time on he occupied the chair earlier graced by Gauss and Dirichlet.

On August 11 Riemann was elected a corresponding member of the Berlin Academy of Sciences in the "physical-mathematics class." The presentation, dated July 4, 1859 and signed by Ernst Eduard Kummer (1810–1893), Carl Wilhelm Borchardt (1817–1880), and Weierstrass, reads in part:

> Prior to the appearance of his most recent work [the "Theory of Abelian functions"] Herr Riemann was almost unknown to mathematicians. This circumstance excuses somewhat the necessity of a more detailed examination of his works as the basis of our presentation. We considered it our duty to turn the attention of the Academy to our colleague whom we recommend not as a young talent who shows great promise, but rather as a fully mature and independent investigator in our area of science, whose progress he has promoted in significant measure.

The papers on which Riemann's election to the Academy was based were his doctoral dissertation and the "Theory of Abelian functions." The paper "On the Hypotheses That Lie at the Foundations of Geometry" was not mentioned at all. Either it remained unknown to Berlin mathematicians or, more probably, it was not considered a serious scientific work. This work also was not mentioned when Riemann was chosen as a foreign member of the Berlin Academy.

In the same year (1859), Riemann received other academic distinctions as well. In November he became a corresponding member of the Bavarian Academy of Sciences (a full member from 1863 on) and a full member of the Göttingen Scientific Society.

The first duty a newly elected correspondent of the Berlin Academy had to fulfill, according to the Academy's Charter, was to send a report on his most recent work. Riemann chose his work on the distribution of prime numbers, "A topic perhaps, which will not be bereft of interest if one recalls that for a prolonged period of time it attracted the attention of Gauss and Dirichlet." The work of which the author spoke so modestly is entitled "On the number of primes less than a given magnitude." It brought forward problems that determined the development of several branches of mathematics for a whole century.

In order to investigate the distribution of prime numbers, Euler had already studied the ζ-function ζ (the notation is Riemann's). Euler had obtained the relation

$$\zeta(s) = \sum_{n=1}^{\infty} \frac{1}{n^s} = \prod (1 - p^{-s})^{-1}. \tag{5.1}$$

(The product extends over all prime numbers and the sum over all positive integers. Euler considered this relation for real s.) It follows immediately from this formula that the set of prime numbers is infinite (the series for $\zeta(1)$ diverges). Mathematicians directed their efforts toward obtaining more precise information about the distribution of primes.

Let us denote the number of primes less than a given number x by $\pi(x)$. There was already a conjecture, apparently enunciated by Euler himself, that as $x \to \infty$

$$\frac{\pi(x)}{\frac{x}{\log x}} \to 1. \tag{5.2}$$

Despite the efforts of such talented mathematicians as Euler, Legendre, and Gauss, this conjecture had not been proved. Gauss, with his characteristic energy and love of computation, even constructed a table of the primes less than three million.

The strongest results before the time of Riemann were obtained by the great Russian mathematician Pafnutii L'vovich Chebyshev (1821–1894) and were published in 1854 in two papers: "Sur la fonction qui détermine la totalité des nombres premiers inférieurs à une limit donnée." (*Mémoires des savants étrangers de l'Académie Impériale Scientifique de St. Pétersbourg*, **VI** (1848), pp. 1–19) and "Mémoire sur les nombres premiers," (*Ibid.*, **VII** (1850), pp. 17–33). Both papers were later reprinted in Liouville's *Journal de Mathêmatiques Pures et Appliquées*, **XVII** (1852), pp. 341–365, 366–390. Chebyshev proved that

$$A_1 < \frac{\pi(x)}{\frac{x}{\log x}} < A_2$$

(where $0.992 < A_1 < 1$ and $1 < A_2 < 1.105$), but he did not prove that the limit exists. To prove the inequality, he used the function $\psi(x)$—the sum of the natural logarithms of all the primes not greater than x.

PAFNUTII L'VOVICH CHEBYSHEV

Riemann seems to have known of Chebyshev's work.[1] Nevertheless, there are no references to Chebyshev in the published text, which may be due both to the broad formulation of the problem and to his essentially different method of solving it. Riemann began his work with the Euler identity (5.1), but he considered the series (5.1) for complex values of s. This was a completely new step in the study of the ζ-function. We shall follow the accepted modern terminology and call it the Riemann ζ-function. To prove the possibility of analytic continuation of ζ to the whole complex plane Riemann introduced a functional equation for the ζ-function.

Just how many important conclusions can be derived from analytic properties of this function can be seen if one looks only at the example of the problem of the

[1]Indirect, but rather convincing evidence for this assertion is given in the book of Edwards (Edwards, H.M. *Riemann's Zeta Function*, New York: Academic Press, 1974). Besides Riemann's close association with Dirichlet, who in turn was personally acquainted with Chebyshev and met him during Chebyshev's visit to Germany in 1852, it turns out that shortly after the publication of his famous papers Chebyshev's name appeared in Riemann's notebooks, which are preserved in the library of the University of Göttingen. A facsimile of the corresponding pages of Riemann's manuscript is printed in the book of Edwards.

distribution of primes. The law of distribution of primes (formula (5.2)) is equivalent to the assertion that the ζ-function of Riemann does not have complex zeros with real part equal to 1: $\zeta(1+it) \neq 0$ when $t \neq 0$. However, most remarkable is the assertion that all zeros of the function $\zeta(s)$, with the exception of trivial ones $(-2, -4, \ldots, -2n, \ldots)$ lie on the straight line $\operatorname{Re} s = 1/2$. This is the famous *Riemann hypothesis*, which remains unproved at the present time.

The majority of results in this work were rigorously substantiated by subsequent generations of mathematicians. In particular, Jacques Hadamard (1865–1963) and Charles de la Vallée-Poussin (1866–1962) proved the validity of the formula for $\pi(x)$. But, as Hadamard himself wrote, "As for the properties for which he gave only a formula, it took me almost three decades before I could prove them, all except one." This last-mentioned property is, in fact, the Riemann hypothesis. Additional material connected with $\zeta(s)$ was found in Riemann's manuscripts preserved in the archives of Göttingen University. Unfortunately, one cannot obtain any sort of proof on the basis of these papers; they give only an idea of the considerations that led Riemann to his hypothesis. Hadamard recalled an assertion contained in Riemann's papers: "These properties are deduced from a representation of it that I was unable to simplify enough to publish."[2] This sentence calls to mind the note made by Pierre Fermat (1601–1665) in connection with his no less famous theorem on the impossibility of solving the equation $x^n + y^n = z^n$ in positive integers when $n > 2$.

It is curious that the analogue of the Riemann hypothesis for ζ-functions defined over finite fields of algebraic numbers, the so-called L-congruence functions and the Artin L-functions, was proved by André Weil (1906–1998) in 1941. Thus one can say that the fields of rational and complex numbers to which we are accustomed have a more complicated structure than other fields. The attempts to prove the Riemann hypothesis undertaken by a number of eminent mathematicians have been very fruitful for the development of analytic number theory. On the one hand it has been possible to reduce many classical problems of number theory to certain assertions about the behavior of the ζ-function, for example the problem of representing a sufficiently large odd number as the sum of three primes, which is the ternary Goldbach problem, solved in 1937 by Ivan Matveevich Vinogradov (1891–1983). On the other hand a number of important results have been obtained toward a proof of the Riemann hypothesis itself. As early as 1914 G.H. Hardy (1877–1947) proved that there exists an infinite set of zeros of $\zeta(s)$ on the line $\operatorname{Re} s = 1/2$. Later Hardy and J.E. Littlewood (1885–1977) obtained an estimate for the number $N_0(T)$ of zeros of $\zeta(s)$, $s = \sigma + it$, on the interval of complex numbers $(1/2) + it$, $0 < t \leq T$, namely $N_0(T) > AT$.

In 1942 the Norwegian mathematician A. Selberg significantly improved the results of Hardy and Littlewood by obtaining the estimate $N_0(T) > AT \ln T$, where A is a sufficiently small constant. Not until 30 years later did the American mathematician N. Levinson succeed in improving Selberg's estimate by proving that $A \sim 1/3$. He thereby proved that at least one-third of the nontrivial zeros of $\zeta(s)$ lie on the critical line.

---

[2]Riemann, B. *Gesammelte Mathematische Werke*, p. 554.

W. WEBER

The paper "Über die Anzahl..." became one of the most famous of Riemann's papers, his only publication on number theory. Another paper, written in 1863 and left unfinished at his death, bears the title "Sullo svolgimento del quoziente di due serie ipergeometriche in frazione continua infinita" ("On the infinite continued-fraction expansion of the quotient of two hypergeometric series"), Riemann considered the seemingly special problem of the convergence of the continued fraction expansions of certain classes of functions, including the hypergeometric functions. But the power of Riemann's talent was so great that even in solving this problem he obtained results that went far beyond the specific purpose he had set. In particular, in this paper he reduced the problem under investigation to the problem of finding asymptotic formulas for multivalued integrals and obtained such formulas using a significant modification of the saddle-point method. A century later the results in this note were used by the British mathematician Alan Baker in his remarkable papers on rational approximations of algebraic numbers.

After Dirichlet's death, in Klein's words, "Riemann may once more have been more strongly influenced by W. Weber,"[3] and he returned to problems of mathemat-

[3]Klein, F. *Development of Mathematics*, p. 237.

ical physics. We will begin with a discussion of the article, "On the propagation of planar air-waves with finite amplitude," published in 1860 in the *Abhandlungen* of the Göttingen Scientific Society. It is devoted to the solution of an important problem on the propagation of sound waves in a medium, given certain relations between pressure and density. This is an exemplary investigation in mathematical physics. At the beginning of the article, Riemann summarized the state of that area of investigation. Discussing the latest articles of James Prescott Joule (1818–1889), W. Thomson (Lord Kelvin), and others on the thermal capacity of gases, it occurred to him that, although known thermal processes are well described by formulas of adiabatic and isothermal expansion, it would be of interest to find solutions of the equations of one-dimensional gas dynamics, given an arbitrary dependence of pressure p on the density of gas ρ. In the abstract to the article, he made a remarkable formulation of the principles that guided him in carrying out the work.

> The present work does not claim to lead to results in experimental research; the author asks that it be considered only as a contribution to the theory of nonlinear partial differential equations. In the theory of linear partial differential equations, by far the most fruitful methods have been developed not by reasoning abstractly on the subject, but rather by studying special physical problems. Likewise, the theory of nonlinear equations can, it seems, achieve the most success if we direct our attention to special problems having physical content with thoroughness and with a consideration of all auxiliary conditions. In fact, the solution of the very special problem that forms the subject of the current paper requires new methods and concepts and leads to results that probably will also play some role in more general problems.[4]

Riemann solved the problem of the propagation of a wave in a homogeneous medium. He derived his equations, analogous to Euler's equations of hydrodynamics, from conservation laws. He obtained a system of equations to describe the motion along the x-axis:

$$\begin{aligned} \frac{\partial u}{\partial t} + u\frac{\partial u}{\partial x} &= -\phi'(\rho)\frac{\partial \rho}{\partial x} \\ \frac{\partial \rho}{\partial t} + \frac{\partial(\rho u)}{\partial x} &= 0, \end{aligned} \tag{5.3}$$

where u is the speed at point x and ρ is the density. The first equation is simply the equation of motion (Newton's second law), and the second is the equation of continuity.

This system of equations is nonlinear. Riemann discovered a remarkable effect that occurs only in nonlinear systems—the existence of "shock wave" solutions (Riemann's term). Physically they indicate that, even though "smooth" initial conditions are given, sharp jumps of pressure and density (breaks in solutions of equations of

[4] Riemann, B. *Gesammelte Mathematische Werke*, p. 176.

type (5.3)) are possible. Riemann gave conditions for the formation of a shock wave (the tipping condition) and found conditions for breaks given the formation of shock waves.

The work of Riemann was developed by the French mathematician Pierre Henri Hugoniot (1851–1887) and the British engineer and physicist William John Macquorn Rankine (1820–1872), who showed how to obtain conditions for jumps from conservation laws. Shock waves raise a number of problems concerning gas and fluid dynamics that have enormous significance in applications. For example, shock waves are formed when high-speed aircraft break the sound barrier, when atomic bombs explode, and so forth.

In this realm of investigation Riemann's work is acknowledged as a classic. In its purely mathematical aspect it is the origin of investigations on general solutions of differential equations of hyperbolic type, out of which arose the theory of distributions.

The fate of another of Riemann's papers, "On the movement of a liquid homogeneous ellipsoid" (1861), is also interesting. It was closely associated with Dirichlet's last work, published soon after his death by Dedekind. Although Riemann himself wrote that

> For a mathematician it is especially attractive to follow along a path whose beginning was posed by this wonderful discovery, completely independently of the question as to the form of heavenly bodies which served as the occasion for these investigations,[5]

it is worth recounting the origin of this problem, which has engaged some of the greatest mathematicians, beginning with Newton.

Newton showed that a slow rotation of the Earth ought to lead to a very slight flattening of it. Newton's arguments were remarkably clever and convincing. Imagine that two boreholes of identical diameter have been drilled in the Earth: one from a point on the equator to the center, the other from the pole to the center. Imagine them filled with a liquid. By the laws of liquid equilibrium, the weight of the column of liquid should be the same in both cases (since the pressure at the center is the same in both directions). However, the gravitational acceleration along the equatorial radius g_{eq} is decreased by centrifugal acceleration g_{c}, while the acceleration along the polar radius g_{p} is not (the Earth rotates about this axis.)

If we assume that the Earth is homogeneous, the accelerations g_{c} and g_{eq} are proportional to the distance from the center of the Earth; therefore their ratio is a constant quantity that can be determined at the surface. We shall denote this ratio by m, the equatorial radius by a, and the polar radius by b. Then the equatorial column of liquid will weigh

$$\frac{a}{2} g_{\mathrm{eq}}(1 - m),$$

[5] *Ibid.*, p. 182.

while the weight of the polar column is

$$\frac{b}{2} g_{\mathrm{p}}.$$

Since these weights are equal, it follows that

$$a g_{\mathrm{eq}}(1 - m) = b g_{\mathrm{p}}. \tag{5.4}$$

For a slightly flattened body

$$\frac{g_{\mathrm{p}}}{g_{\mathrm{eq}}} = 1 + \frac{1}{5}e + O(e^2), \quad e = 1 - \frac{b}{a}.$$

From these two formulas one obtains the amount of flattening:

$$e = \frac{5}{4}m.$$

In Newton's time the quantity m was known ($m = 1/290$), and from it Newton obtained the amount of flattening: $e \sim 1/230$.

Newton's conclusion contradicted astronomical facts of the time and "two generations of the best astronomical observers formed in the school of the Cassinis[6] had struggled in vain against the authority and the reasoning of Newton."[7] The juxtaposition of the views of Newton and Jacques Cassini (1677–1756) are beautifully illustrated in an old drawing (Fig. 5.1). Nonetheless, it required geodetic measurements, conducted in Lapland in 1738 by Clairaut and Pierre Louis Moreau de Maupertuis (1698–1759), to confirm definitively the phenomenon of the flattening of the Earth's surface at the poles. As Isaac Todhunter (1820–1884) wrote:

> The success of the arctic expedition in large measure ought to be attributed to the artistry and energy of Maupertuis, and his name became widely known. In prints of the time he is shown in the costume of a Lapland Hercules: a fur hat pulled over the eyes; in one hand he held a cudgel, and in the other he gripped a globe.

Voltaire, a friend of Maupertuis, warmly congratulated him at the time for flattening both the poles and the adherents of Cassini. Later Maupertuis and Voltaire got into a tragicomic polemic, and Voltaire wrote,

> "Vous avez confirmé dans les lieux pleins d'ennui
> Ce que Newton a connu sans sortir de chez lui."
>
> ("You went out into the wilderness to confirm
> What Newton knew without leaving home.")[8]

---

[6]The Cassinis were a family of French astronomers. The most famous were Gian Domenico (1625–1712), Jacques (1677–1756, son of Gian Domenico), and César François (1714–1784, grandson of Gian Domenico).

[7]Todhunter, I. *History of the Mathematical Theories of Attraction and the Figure of the Earth from the Time of Newton to that of Laplace*, London: Constable, 1873. Reprinted by Dover, New York, 1962.

[8]Quoted in Chandrasekhar, S. *Ellipsoidal Figures of Equilibrium*, New York: Yale University Press, 1969.

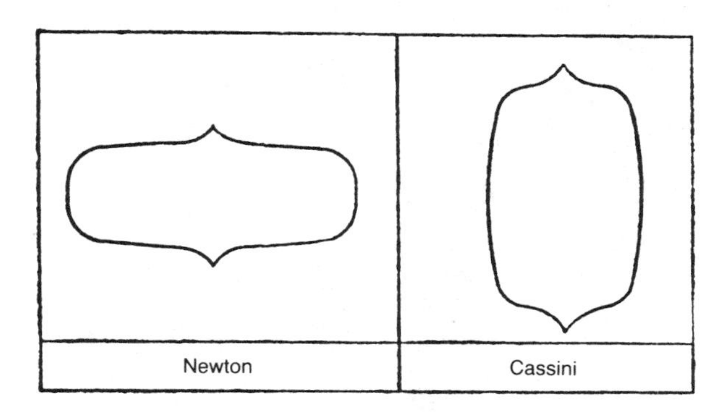

Figure 5.1: The views of Newton and Cassini.

In 1742 Colin Maclaurin (1698–1746) considered the general problem of the equilibrium of a rapidly rotating homogeneous body. He showed that such figures will be oblate spheroids. Laplace, d'Alembert and, even earlier, Thomas Simpson (1710–1761) the discoverer of the well-known Simpson rule, noticed a remarkable property of Maclaurin's spheroids. It does not follow that the spheroid is nearly a sphere when the angular velocity is small. It turns out that there exist two solutions; the first one is indeed nearly a sphere, but the second is nearly a very oblate ellipsoid.

For almost a century it was thought that Maclaurin had found a complete solution to the problem of figures of equilibrium of rotating homogeneous bodies. Only in 1834 did Jacobi point out that arbitrary ellipsoids might be permissible figures of equilibrium. The figures of equilibrium (stationary figures) found by Maclaurin and Jacobi possess one important special feature: they disintegrate if their angular velocity of rotation exceeds a certain magnitude. Dirichlet posed the following problem: *Determine the motion of a gravitating body that has the shape of an ellipsoid at each instant of time, when the coordinates of the particles are linear functions of their initial values.* Dirichlet himself considered the special case of a spheroid. Dedekind, who prepared Dirichlet's work for publication, found yet another class of solutions.

A complete solution of Dirichlet's problem of stationary figures was given only by Riemann. He showed that, given a linear dependence of the velocity field on coordinates of the most general type of motion under which the ellipsoidal shape of the figure of equilibrium is preserved, there is a superposition of uniform rotation and internal movements with uniform vorticity of the liquid. Riemann's ellipsoids include the classical ellipsoids of Maclaurin, Jacobi, and Dedekind and, in addition, three new classes of figures. Riemann also touched on the problem of the stability of figures of equilibrium. In doing this he applied his beloved variational principle; but, as the American physicist N. Lebovitz has recently pointed out, he made an error in this part.

The question of the stability of such figures, which is quite important for the theory of the origin of planets, was investigated subsequently in the classic works of

ALEKSANDR MIKHAILOVICH LYAPUNOV

A.M. Lyapunov (1857–1918) and Poincaré. At present there is increasing interest in these fundamental results in connection with problems of astrophysics (for example, in the investigation of the equilibrium state of stars, for example, the structure and stability of white dwarfs and neutron stars). For the majority of stars one can apply Newton's theory of attraction. In the framework of Newton's theory the well-known astrophysicist Subramanyan Chandrasekhar (1911–1995) found, for example, an upper bound on the mass of white dwarfs, below which they remain stable. It would be extremely interesting to consider the problem of equilibrium in relativistic hydrodynamics, although its potential applications, for example, to the study of superdense stars, are unclear; aside from purely dynamic properties, the physical processes themselves are important in stars. Riemann's note, "Equilibrium of electricity on circular cylinders," evidently dates to this same period. The problem of the distribution of electrical charge in cylindrical conductors leads to the purely mathematical problem of solving Laplace's equation in a simply connected domain with prescribed boundary conditions. Here, for the first time, automorphic functions arise.

All these remarkable works of Riemann were carried out in three years. At the same time he became acquainted with the greatest mathematicians of the world. In 1859, soon after being elected into the Academy, he came to Berlin, where Kummer, Weierstrass, and Kronecker received him warmly. Weierstrass pointed out to him that the generalization of the well-known theorem of Jacobi that a single-valued analytic function on the complex plane cannot have more than two periods is of great

HENRI POINCARÉ

LEOPOLD KRONECKER

CHARLES HERMITE

significance for the theory of Abelian functions. Somewhat later Riemann reported in a letter to Weierstrass his own proof of an analogous multidimensional theorem: *A single-valued analytic function of n-complex variables cannot have more than $2n$ periods.*

Riemann took another trip in March of 1860 which brought him great pleasure. He visited Paris where he met with the most eminent French mathematicians: Charles Hermite (1822–1901), Joseph Bertrand (1822–1900), and Victor Puiseux (1820–1883). He also established friendly relations with the inseparable pair Charles Briot (1817–1882) and Jean-Claude Bouquet (1819–1885), well known mathematicians and authors of the first textbook on analytic functions.

Chapter 6

Last Years

"I am always saddened when talented people die because earth needs them more than heaven does."

GEORG LICHTENBERG

IN July 1862 Riemann married Elise Koch, a friend of his sister. This marriage brightened the last years of his life. In the fall of 1862 he caught a serious cold and tuberculosis set in. Riemann spent his last years almost entirely in Italy. Thanks to a petition of Weber and Sartorius von Waltershausen, known for his work on Mt. Etna, he received government subsidies on three occasions for recuperation. He returned to Göttingen, but the climate was too severe for him, and he did not attempt to give lectures. Riemann's last trip to Italy coincided with the beginning of the Seven Weeks' War between Hannover and Prussia. He arrived in Italy on June 28, 1866, the day Hannover capitulated.

In Italy Riemann had many friends among the most important Italian mathematicians, including the acquaintances he made while still in Göttingen, E. Betti, and also E. Beltrami. Riemann's influence left a strong imprint on the Italian school of mathematics.

When his illness was in remission, he continued to work. His final work, "Mechanik des Ohres" ("Mechanics of the ear"), published posthumously, was accompanied by the following editorial comment:

> During the last months of his life, this outstanding mathematician, who was taken from our school and from science by an early death, studied the theory of hearing inspired by Helmholtz's new theory of auditory sensations.[1]

It is known that Riemann was also interested in the nature of sight.

Riemann spent the last months of his life in the little village of Selasca on Lake Maggiore together with his wife and three-year-old daughter, Ida.

[1] Riemann, B. *Gesammelte Mathematische Werke*, p. 338.

At the end of his life Riemann received recognition for his works. The Berlin Academy, which was the first to have noticed Riemann, elected him a foreign member at the beginning of 1866. From the presentation signed by Kronecker, Borchardt, and Weierstrass, it is clear how immeasurably the appraisal of Riemann's merits had grown:

> We confine ourselves to the results of only three of Riemann's works[2] because they are already adequate for an appraisal of his significance. At the same time, however, we must emphasize that his methods deserve the admiration of specialists no less than his results and that a treasure trove of the most important and fruitful observations are to be found in the individual conclusions and propositions made in these works.
>
> After we have clearly described not only the rare talent of Herr Riemann but also the exceptional place that his name doubtless ought to occupy in the history of science, there is every reason to award him the Academy's highest distinction—election as a foreign member.

In March 1866 Riemann was elected a foreign member of the Paris Academy of Science, and on June 14, 1866, a month prior to his death, a member of the London Royal Society. "His strength declined rapidly, and he himself felt that his end was near," wrote Dedekind in his biography. "But still, the day before his death, resting under a fig tree, his soul filled with joy at the glorious landscape, he worked on his final paper which, unfortunately, was left unfinished."[3]

Riemann died on July 20, 1866, his mind clear to the last second. His last words to his wife were, "Kiss our child." He was buried in the nearby village Biganzola. His tombstone bears the following epitaph: "Here lies in God Georg Friedrich Bernhard Riemann—Göttingen professor, born in Breselenz, September 17, 1826, died in Selasca, July 20, 1866. Denen die Gott lieben müssen alle Dinge zum Besten dienen."[4] ("All things work together for good to them that love God." [*Rom.* 8:28])

---

[2]Borchardt, Kronecker and Weierstrass mentioned the following papers: "Über die Anzahl der Primzahlen unter einer gegebenen Grösse" (On the number of primes less than a given quantity), "Über die Fortpflanzung ebener Luftwellen von endlicher Schwingungsweite." ("On the propagation of planar air waves of finite amplitude"), "Ein Beitrag zu den Untersuchungen über die Bewegung eines flüssigen gleichartigen Ellipsoid" ("A contribution to the study of the motion of a liquid ellipsoid of revolution").

[3]Riemann, B. *Gesammelte Mathematische Werke*, p. 558.

[4]*Ibid.*, p. 558.

Chapter 7

Posthumous Fate

"The discovery of a new truth in itself is the greatest happiness; recognition can add almost nothing to that."

FRANZ NEUMANN

WHILE he lived, Riemann's work did not procure their author the influence he had a right to claim. This is especially true of the works that today are considered perhaps his main contribution to science: the theory of Abelian integrals, Riemann surfaces, and Riemannian geometry. There are many objective and subjective reasons that explain this circumstance. His views on geometry were, of course, completely novel for a wide circle of mathematicians. One must not forget that the very idea of non-Euclidean geometry was accepted only with difficulty, even eliciting wild fury from a majority of philosophers. For example, here is what E.K. Dühring wrote in his essay, "Kritische Geschichte der allgemeinen Prinzipien der Mechanik" (A critical history of the general principles of mechanics), which earned the Benecke Prize in 1872 from the philosophical faculty of Göttingen University:

> Thus the late Göttingen mathematics professor, Riemann, who—with his lack of originality except for Gaussian self-mystification—was also led astray by Herbart's philosophistry, wrote (in his paper, "On the hypotheses that lie at the foundations of geometry, *Göttinger Abhandlungen*, Vol. 13, 1868): "But it seems that the empirical concepts on which the spatial definitions of the physical universe are based, the concept of a rigid body and of a light ray, are no longer valid on the infinitesimal level. Thus, it is permissible to think that physical relations in space in the infinitely small do not correspond to the axioms of geometry; and, in fact, this may be assumed if it leads to a simpler explanation of phenomena." It is not surprising that the somewhat unclearly philosophizing physiological professor of physics, H. Helmholtz, also could not pass up the opportunity to meddle in these investigations. In the article, "On the facts that lie at the

> foundations of geometry,"[1] he commented upon this curious absurdity in a favorable sense.

Years passed before the ideas of Riemann and Helmholtz found their mathematical realization in the works of Poincaré and Einstein. The works on the theory of Abelian functions—which were Riemann's most brilliant in terms of the ideas they contained—were written in Riemann's characteristically intuitive manner; in places they were based on unproved assertions, and they did not satisfy the standards of rigor that were established at the time in mathematics. As an example of the extreme incomprehension of the value of Riemann's work, we again refer to the work of Dühring:

> Evidence of the completely dependent character of Riemann's work on Abelian functions is that in it the same method of intuitive presentation, in the same completely arbitrary form, is taken and extended simply by faith in the authority of the teacher [Gauss].

Today Dühring's article is considered a curio, but one must remember that in his time he was a well-known and influential philosopher.[2] He was known for his pathological antisemitism and for his acrimonious dispute with Helmholtz, but other qualified mathematicians also showed no clear understanding of Riemann's work.

Weierstrass joined the fray with a criticism of the Dirichlet principle, on which Riemann's work was based, but Riemann's ideas gradually gained full recognition. After Hilbert proved the Dirichlet principle, all Riemann's arguments acquired a firm basis. Riemann's greatness as a mathematician lies in the fact that almost all of his works proved to be not an ending but rather the beginning of new, productive research. One can cite the theory of automorphic functions, the Atiyah–Singer theorem on the index of differential operators on manifolds of arbitrary dimensions (a generalization of the Riemann–Roch theorem), the problem of moduli of complex manifolds, the theory of ζ-functions on algebraic varieties, Selberg's trace formula in the theory of discrete groups, and much more.

Riemann's successes in so many areas of physics and mathematics depend on his universal approach to natural phenomena and his unusual flair for comprehending connections between apparently disparate phenomena. In his philosophical reflections he wrote, "My main work is in the area of a new understanding of well-known laws of Nature."

Here it is perhaps appropriate to note the differences between two great mathematicians, Riemann and Weierstrass, whose names often stand side by side in contemporary mathematics. Weierstrass' rigorous approach, in contrast to Riemann's intuitionist approach, consisted of a precise and sequential process of reasoning. Weierstrass adhered to this approach in his articles and lectures. His concepts of rigor

---

[1] *Abhandlungen der Königlichen Gesellschaft der Wissenschaften zu Göttingen*, June 1868.

[2] A critique of many of Dühring's ideas can be found in Friedrich Engels', "Anti-Dühring." For rather comical reasons Dühring's name (but, of course, not his work) was well-known in the Soviet Union, where every student was required to read Engels' work.

DAVID HILBERT

later became the standard in mathematical works. Weierstrass is considered a classic example of a pure mathematician, and his pronouncement on the connection of mathematics with its applications is, therefore, all the more interesting:

> Between mathematics and the natural sciences, deeper mutual relations ought to be established than those hold when, for example, the physicist sees in mathematics only an auxiliary, though necessary, discipline, while mathematicians regard the questions posed by physicists only as a bountiful collection of examples for their methods... To the question, Can one really obtain anything directly applicable from those abstract theories with which today's contemporary mathematicians occupy themselves?, I can answer that Greek mathematicians studied the properties of conic sections in a purely theoretical way long before the time when anyone could foresee that these curves represent the paths along which the planets move. I believe that many more functions with such properties will be found; for example, the well-known θ-functions of Jacobi make it possible, on the one hand, to find the number of squares into which any given number decomposes, thereby making it possible to rectify an arc

> of an ellipse, and, on the other hand, they make it possible to find the true law of the oscillations of a pendulum.

The clear difference between Weierstrass and Riemann can be seen not only in the difficult concrete physics problems they solved but in the effort to build theories explaining natural phenomena.

These two contrasting views on the aims of mathematics can be traced at all stages of the development of science. Jacobi presented them in clear form:

> Monsieur Fourier held the opinion that the main aim of mathematics is its social utility and the explanation of phenomena of Nature; but as a philosopher, he ought to have known that the single goal of science is to bolster the courage of human reason, and therefore any sort of question in the theory of numbers has no less value than a question about the system of the world.

In the actual evolution of science, by far the most fruitful approach is to accept the coexistence of both points of view. Niels Bohr (1885–1962) would call both approaches great truths. (A great truth is a truth whose negation is also a great truth.) We see a confirmation of this in the works of Jacobi on mechanics and of Riemann on the theory of numbers. Nonetheless, the history of mathematics shows periods when one of these tendencies is more prevalent than the other.

A typical "intuitionist," Klein, who did much to develop Riemann's ideas, gave an appreciation, in humorous form, of the mathematics of the end of the nineteenth century:

> Mathematics in our day reminds me of major small-arms production in peacetime. The shop window is filled with models that delight the expert by their cleverness and their artful and captivating execution. Properly speaking, the origin and significance of these things—that is, their ability to shoot and hit the enemy, recedes in one's consciousness and is even completely forgotten.

Just such a situation arose in the 1940s and 1950s. It was very aptly characterized by the well-known American theoretical physicist and mathematician Freeman Dyson: "The marriage between mathematics and physics, which was so fruitful in past centuries, recently ended in divorce." Various causes have brought about this situation. Now, of course, it is absolutely impossible to imagine a mathematician working in an experimental physics laboratory. The time of Riemann will not come again, but recently we have seen renewed interest in physics among "pure" mathematicians. This renewed interest is now bearing its first fruits, and in precisely those areas of mathematics in which Riemann worked.

In this situation completely new connections among various subjects that occupied Riemann have arisen in remarkable fashion. For example, in the theory of nonlinear waves (as well as for other classes of equations, i.e., the so-called integrable evolutionary systems), the Riemann θ-function arises in the search for periodic solutions.

The Selberg trace formula, which has roots in Riemann's work on the theory of the ζ-function and the works of Klein and Poincaré on the theory of automorphic functions, makes it possible to exhibit a connection between the distribution of the eigenvalues of the Laplacian on Riemann surfaces and the geometric and topological characteristics of the surface. Interest in this circle of questions has increased sharply in connection with the problems of quantizing classical dynamic systems with stochastic behavior ("quantum chaos").

The theory of gravitation, relying on Riemannian geometry, is merging with the theory of strong interactions (via the theory of gauge fields). A quantum theory of gravitation is coming into existence, in which vigorous use is being made of the methods of algebraic geometry, which originated in Riemann's work on Abelian functions. One can now speak boldly of a "renaissance" in the relation between mathematics and physics. It suffices merely to list some of the brilliant achievements of modern mathematics that owe their existence to physics/mathematics "inbreeding": the construction of different smooth structures on 4-dimensional Euclidean space (fake R^4), the discovery of new knot invariants, the creation of the theory of infinite-dimensional Lie algebras and quantum groups, and the effective description of the space of moduli of Riemann surfaces.

More than a hundred years have passed since Riemann's death. Mathematics has been enriched in a major way by new ideas and results. Cantor's theory of sets has transformed its face and the degree of abstraction has reached extraordinary levels—one has only to recall the theory of categories and functors, the theory of formal schemes, motives, the l-adic cohomology of Grothendieck, and so forth. It is remarkable that from the height of these theories Riemannian concepts that seemed completely mysterious and unrigorous to his contemporaries have received a very adequate description in the language of modern algebraic geometry and topology.

A mathematical idea is fruitful if it makes progress possible on complex concrete problems left by preceding generations. Now, as before, the Riemann hypothesis poses a challenge to contemporary active mathematicians. One cannot doubt that Riemann's works will interest not only historians of science but also mathematicians for many years to come.

Bibliography to Part I

1. Anosov, D., Bolibruch, A., *The Riemann–Hilbert Problem*, Vieweg, Braunschweig, 1994.

2. Chandrasekharan, K., *Introduction to Analytic Number Theory*, Springer-Verlag, New York, 1968.

3. Farkas, H. Kra, I., *Riemann Surfaces*, 2nd ed, Springer-Verlag, New York, Heidelberg, Berlin, 1996.

4. Gray, J., *Linear Differential Equations and Group Theory from Riemann to Poincaré*, Birkhäuser Boston, 1986.

5. Griffits, P., Harris, J., *Principles of Algebraic Geometry*, John Wiley & Sons, New York, 1978.

6. Klein, F., *Vorlesungen über die Entwicklung der Mathematik im 19. Jahrhundert.* Berlin: Springer-Verlag, 1926–1927. (Translation of Vol. I: *Development of Mathematics in the 19th Century.* Trans. M. Ackerman., Math-Sci Press, Brookline, MA, 1979.

7. Laugwitz, D., *Bernhard Riemann, 1826–1866, Wendepunkte in der Auffassung der Mathematik*, Birkhäuser Basel, 1996.

8. Laugwitz, D., *Bernhard Riemann, 1826–1866, Turning Points in the Conception of Mathematics*, Translated by Abe Shenitzer, Birkhäuser Boston, 1999.

9. Misner, C., Thorne, K., Wheeler, J., *Gravitation*, W.H. Freeman, New York, 1973.

10. Riemann, B., *Gesammelte Mathematische Werke* (R. Narasimhan, ed.), Springer-Verlag, New York, Heidelberg, Berlin, 1990.

11. Springer, G., *Introduction to Riemann Surfaces*, Addison–Wesley, Reading, MA, 1957; 2nd ed., Chelsea, New York, 1981.

12. Willmore, T.J., *Riemannian Geometry*, Clarendon Press, Oxford, 1995.

Part II

Topological Themes in Contemporary Physics

IT is far from accidental that topological and differential geometric ideas have emerged almost simultaneously in such seemingly disparate realms of physics as the theory of elementary particles and structural transformations in liquid crystals, quantum gravity and superfluid properties of helium.

The application of these branches of mathematics is connected with basic unifying processes in theoretical physics, in particular with the recognition of the universality of such concepts as gauge invariance, hidden symmetry and the like. Therefore, the effective use of topological methods in this varied circle of questions is not surprising.

Just as the methods of the theory of groups became an indispensable tool in the work of theoreticians after the clarification of the fundamental connection between conservation laws and symmetry principles, so topological techniques are necessary for studying the global properties of field theory and condensed matter. Finally, the effectiveness of topological applications is attained in combination with other methods, especially analytic ones.

The connection of topology with physics is no passing interlude but rather represents a length affair. Readers who are interested in the contemporary state of the basic sciences may wish to learn about the significant events of the recent past in somewhat greater detail.

I have tried to make the exposition accessible to readers with a modest background in physics and mathematics. It was impossible to avoid a certain fragmentation and slant toward mathematics, however, considering the brevity of this book and the variety of themes discussed.

Chapter 8

Introduction

"Geometry is the art of good reasoning from poorly drawn figures."

ANONYMOUS[1]

TOPOLOGY is probably the youngest of the classical branches of mathematics. In contrast to algebra, geometry, and number theory, whose genealogies date to prehistoric times, topology (or as it was earlier called, *analysis situs*, that is, analysis of position, did not appear until the seventeenth century).

In 1679 Leibniz published his famous book *Characteristica Geometrica*, in which (in modern terms) he tried to study the topological rather than the metric characteristics of properties of figures. He wrote that, aside from the coordinate representation of figures, "we are in need of another analysis, purely geometric or linear, which also defines the position (situs), as algebra defines magnitude." It is interesting to note that Leibniz tried to interest Christiaan Huygens (1629–1695) in his work, but the latter showed little enthusiasm. This was the first (albeit unsuccessful) attempt to interest a physicist in topology.

Eighteenth-century mathematicians showed little interest in topology, with the exception of Euler, whose genius comprehended all of mathematics. Euler obtained two purely topological results which played an important role in the development of topology. The first of these—a proof of a classical theorem—is Euler's theorem on polyhedra ($V - E + F = 2$, where V is the number of vertices, E the number of edges, and F the number of faces); the second is the solution of the problem of the seven bridges of Königsberg. The latter laid the basis for the study of the topology of closed curves.

It is instructive to recall the formulation of this problem, which shows that answers to questions that appear useless at first glance can give rise to serious mathematical theories. In the city of Königsberg there were seven bridges across the River Pregel which also connected two islands (Fig. 8.1). The question posed was the following: *Is it possible, departing from any one place on one shore, to trace a path crossing*

[1] Anonymous, quoted by Henri Poincaré in: *Dernières Pensées*, Flammarion, Paris, 1913.

LEONHARD EULER

each bridge exactly once and return to the starting point? For the inhabitants of Königsberg, this problem had more than abstract interest; all seven bridges charged tolls.

In 1735 Euler found a remarkable general solution of this problem. He reduced it to the following question: could one draw the given closed curve without lifting the pencil from the paper and without crossing one and the same line twice? A graph with this property is said to be *Eulerian*. Euler obtained a beautiful constructive criterion for identifying such graphs. A graph Γ is Eulerian if and only if it is connected and the number of vertices of odd degree is zero or two. (The degree of a vertex is the number of edges it belongs to. The number of vertices of odd degree in any graph is even.) Thus, in this very special case, the criteria established by Euler contained the negative solution of the Königsberg bridge problem.

In the nineteenth century topology developed rather slowly until the appearance of the outstanding work of Riemann on algebraic functions. Work in this field was concentrated in Berlin and Göttingen, where August Ferdinand Möbius (1790–1868) and Listing were working. The term *topology* may have appeared for the first time in the work of Listing, a student of Gauss and (incidentally) a future professor of physics at Göttingen. In 1848 he published a book, *Vorstudien zur Topologie*, devoted primarily to a theory of knots. The eighty pages of this valuable book contain discussions of a number of concepts needed for the appearance of topology as an

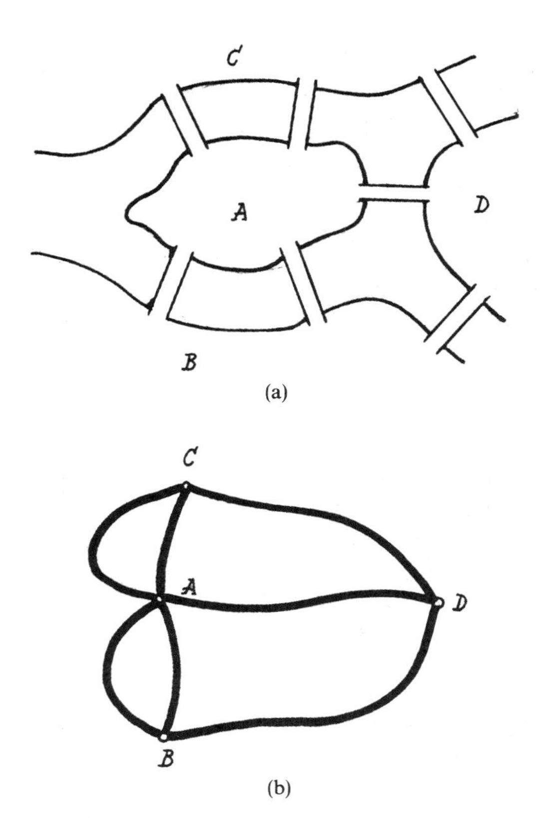

Figure 8.1: The seven bridges of Königsberg. (a) Seven bridges across the Pregel River. (b) An equivalent graph.

independent subject. The main topological topic of the book is the theory of knots and links. Listing came very close to a classification of knots by means of generating relations. The method was later developed by W. Wirtinger (the *Wirtinger presentation* of a knot). Of especial interest to the modern reader are the motivations of the various concepts and the examples, which Listing took from biology (twisted lines in single-shelled snails), botany (hops and the scales of pine cones), and astronomy (the relative positions of the planetary orbits). Listing justified the introduction of the new term *topology* by saying that the phrase *geometry of position* (*geometria situs*) was already being used in a different sense in projective geometry. The essence of his definition of topology is still valid today. "Topology," he wrote, "is the study of the modal relations of spatial figures and the laws of connectivity, mutual position, and ordering of points, lines, surfaces, and solids and their parts independently of measure and magnitude relations." For a long time, however, the term *geometria situs* and the still more widespread term *analysis situs* continued to compete with *topology*, until they were definitively displaced in the 1920's.

The nineteenth century produced several other prominent mathematicians who obtained interesting topological results. Notable examples were Clifford, Klein, Möbius, and Bonnet; several other names could be added. Among the latter were Peter Guthrie Tait, a British mathematician and physicist, a close friend of W. Thomson (Lord Kelvin). His research in topology involved knot theory, in which he obtained a number of first-rate results, including the complete classification of knots having knottedness of order at most eight. (The *knottedness* of a curve C is the number of double points obtained when C is projected onto a plane in general position.) He stated a number of very deep conjectures on the structure of knots, some of which were settled only very recently, after the discovery of new polynomial invariants of knots and links.

It is quite remarkable that Tait was led to his study of knots by reflecting on a theory of atomic structure proposed by Kelvin. Kelvin had invented a rather exotic theory based on the idea that an atom is a clump of tangled vortex lines. In the context of this theory the stability of atoms was explained as the impossibility of untying a nontrivial knot. The spectral characteristics of atoms resulted from the fluctuations of the vortex lines. It is curious that this approach resonates with the modern concepts of string theory. This seductive theory failed rather quickly when it did not survive experimental tests, but Tait's mathematical results remained. This beautiful example of the interaction of physics and mathematics shows the difference in the significance of results in the two subjects. Even the most elegant and beautiful physical theory may disappear without a trace if not confirmed by experiment, while, as a rule, a theorem, once proved, remains in mathematics forever. Like all general assertions, this thesis may provoke some objection; but I hope the reader understands what is meant.

In the nineteenth century several other individual results were obtained which can be classified as topology, but there is nonetheless complete justice in Poincaré's assessment: "After Riemann came Betti, who introduced several fundamental concepts, but after Betti no one else followed." The decisive step was taken by Poincaré himself. In 1895 his paper "Analysis situs" appeared in the *Journal de l'Ecole Polytechnique*—an excellent gift to the school on its hundredth anniversary from one of its alumni. By general agreement among the best mathematicians of the world, it was this work of Poincaré that established topology as an independent branch of mathematics. In this article and its five appendices, Poincaré formulated the basic concepts of the new field. His ideas and results, in fact, defined the future development of topology.

In this first article he underlined the importance of applying topological methods to solve various problems of mathematics. He wrote, "It is easy to see that a generalized *analysis situs* would make it possible to investigate equations of higher orders and, in particular, the equations of celestial mechanics."[2] His foresight has been brilliantly vindicated. Topology has become the basis of the contemporary qualitative theory of differential equations. However, the classical realms of mathematics and mechanics are far from exhausting the possibilities of applying topology. In recent years it has become an organic part of physics in such actively developing fields as

[2] *Ibid.*

the theory of gravitation, quantum field theory, and solid state physics. The primary reason for this is that many concepts in physics and topology are surprisingly close to each other, although formulated in different languages. We will try to follow this connection, illustrating well-known topological concepts with several examples from physics.

Chapter 9

Topological Structures

TOPOLOGY studies those characteristics of figures which are preserved under a certain class of continuous transformations. Imagine two figures, a square and a circular disk, made of rubber. Deformations can convert the square into the disk, but without tearing the figure it is impossible to convert the disk by any deformation into an annulus. In topology, this intuitively obvious distinction is formalized. Two figures which can be transformed into one other by continuous deformations without cutting and pasting are called *homeomorphic*. For example, the totality of sides of any polygon is homeomorphic to a circle, but a circle is not homeomorphic to a straight-line segment; a sphere is homeomorphic to a closed cylinder but not to a torus, and so on.

Poincaré graphically explains the essence of topological structures in his profound *Dernières Pensées:*

> Imagine any sort of model and a copy of it done by an awkward artist: the proportions are altered, lines drawn by a trembling hand are subject to excessive deviation and go off in unexpected directions. From the point of view of metric or even projective geometry these figures are not equivalent, but they appear as such from the point of view of geometry of position [that is, topology].

Of course, Poincaré did not foresee the appearance of abstract artists who saw nothing wrong with disturbing the topology of the model.

The definition of a homeomorphism includes two conditions: continuous and one-to-one correspondence between the points of two figures. The relation between the two properties has fundamental significance for defining such a paramount concept as the dimension of space. Georg Cantor, the founder of the theory of sets, deserves the credit for developing a full logical basis for the concept of dimensionality. From his theory it follows that without enlisting continuity properties it is impossible to define the dimension of a space. For example, one can establish a one-to-one correspondence between points of a straight line, of a plane, and, in general, of the space of any number of dimensions. The concept of dimension acquires precise meaning if one superimposes the additional condition of continuity. The Dutch mathematician L.E.J. Brouwer proved a basic theorem: *There is no one-to-one and continuous mapping with a continuous inverse (homeomorphism) of the space M^n to M^m (given that $n \neq m$).* This theorem led topologists to study the general characteristics of topological spaces. A whole school of set-theoretic topology arose and grew especially rapidly in the 1920s.

The other branch of topology that owes its origins to the first works of Poincaré received the name of *algebraic* or *combinatorial* topology. Algebraic topology studies properties of a narrower class of spaces—basically the classical objects of mathematics: spaces given by systems of algebraic and functional equations, surfaces lying in Euclidean space, and other sets which in mathematics are called *manifolds*. Examining the narrower class of spaces permits deeper penetration into their structure.

At present, most of the applications that have been found are only for the methods of algebraic topology. In the remainder of this book the term, "topology" will signify algebraic topology.

One of the basic tasks of topology is to learn to distinguish nonhomeomorphic figures. To this end one introduces the class of invariant quantities that do not change under homeomorphic transformations of a given figure. The study of the invariance of topological spaces is connected with the solution of a whole series of complex questions: Can one describe a class of invariants of a given manifold? Is there a set of integral invariants that fully characterizes the topological type of a manifold? and so forth. The possibility of ascertaining the topological type of a manifold given a set of integral invariants, is especially interesting. In essence, we are talking about a task very close to physics—to characterize a particle, given its special parameters, for example, spin, charge, mass, etc. (Integral invariants are a sort of "quantum numbers" of a manifold.)

Among such tasks is the classification of two-dimensional surfaces. From this example one can appreciate the beauty and nontriviality of topological methods. There exist two types of surfaces: orientable and nonorientable. A surface is *orientable* if a normal vector returns to its original direction when it is transported over any closed curve on the surface; otherwise the surface is *nonorientable* (Fig. 9.1).

All closed orientable surfaces (without boundary) can be constructed using a special technique. Take a sphere, cut a hole in it, and paste a handle to the hole. A "handle" is a figure homeomorphic to a torus with a hole cut out (Fig. 9.2). Thus we get a figure homeomorphic to a torus. One can obtain all other orientable surfaces by successively cutting holes in the sphere and pasting handles to them. Such a sphere with "handles" will be a model of an arbitrary orientable surface. The number of handles (a topological invariant of the surface) defines what is called the *genus* of the surface. Two surfaces with a different number of handles are nonhomeomorphic (Fig. 9.2).

The genus of the surface is associated with another important number—the Euler characteristic of the surface. Euler's theorem on polyhedra homeomorphic to a sphere admits generalization to an arbitrary surface. Let us draw a grid on an arbitrary two-dimensional surface, "chopping" it into pieces homeomorphic to a disk. Let us denote

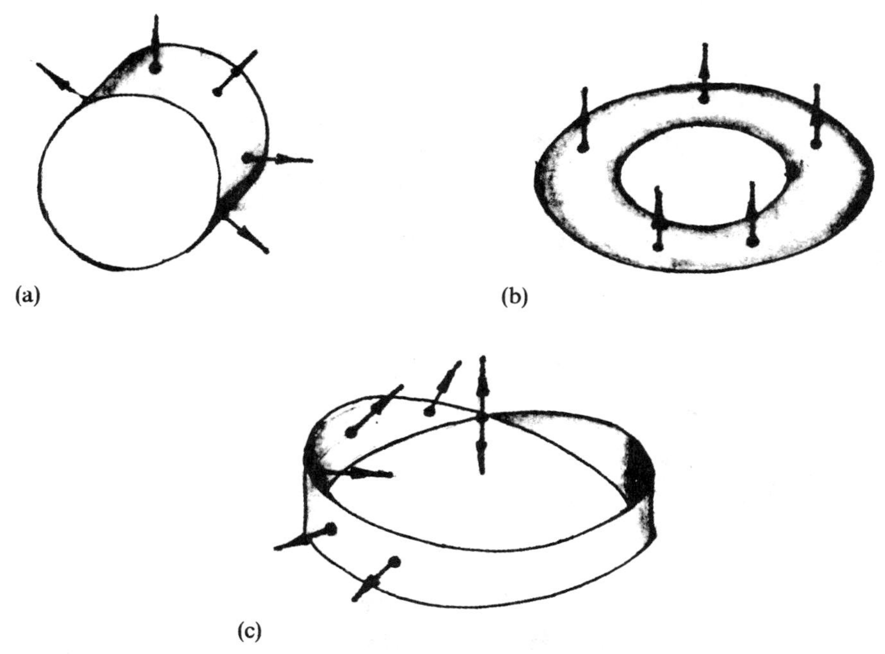

Figure 9.1: Orientable and nonorientable surfaces. (a) A cylinder—an orientable surface with a boundary (two circles). (b) A torus—a closed orientable surface. (c) A Möbius strip—a nonorientable surface with a boundary homeomorphic to a circle. A Möbius strip is a one-sided surface; there exists a path on it such that, if one traverses the path, the normal direction to the surface reverses direction.

the number of pieces of this net by F, and the number of the vertices and the number of the edges by V and E, respectively. The quantity χ defined by $\chi = V - E + F$ is called the *Euler number* or the *Euler characteristic*. This number is a topological invariant of the surface and does not depend on the choice of the net. The Euler characteristic is expressed in terms of the genus of the surface by the equation $\chi = 2 - 2p$, where p is the number of "handles" or the type of the surface.

Nonorientable surfaces can also be obtained by a simple construction. Here the Möbius strip plays the role of "handle." The Möbius strip, the first example of a nonorientable surface, was discovered by Möbius and Listing independently in 1858. It is easy to construct. Take a rectangular strip, give it a half-twist, and glue the ends together (Fig. 9.3a). The Möbius strip is not a closed surface. Its edge is homeomorphic to the circle. An arbitrary closed nonorientable surface, N_q, can be represented as a sphere in which a certain number of holes are cut out and a Möbius strip is glued to each hole. In contrast to orientable surfaces, where the process of gluing is feasible in real three-dimensional space, nonorientable figures cannot be constructed without self-intersection in three-dimensional space (Fig. 9.3). Closed nonorientable surfaces can be imbedded without self-intersection only in four-dimensional space. Nonethe-

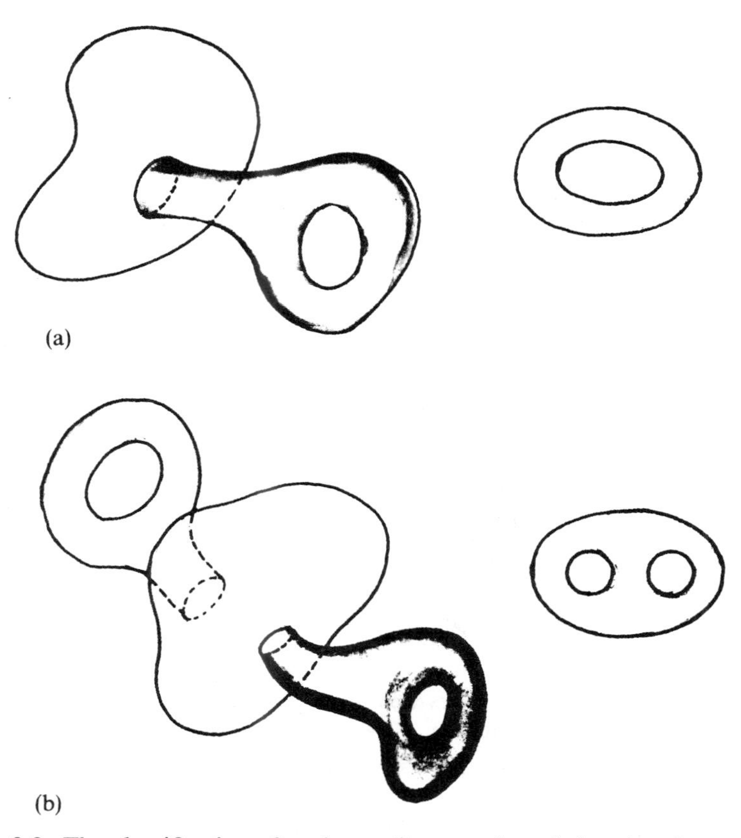

Figure 9.2: The classification of surfaces. Construction of closed surfaces of genus one and two. (a) Conversion from a sphere to a torus. (b) Conversion from a sphere to a doughnut with two holes.

less, from a topological point of view, nonorientable surfaces are homeomorphic to the surfaces N_q. The Euler characteristic of the Möbius strip is equal to zero; cutting a hole out of a surface (the removal of one face) reduces its Euler characteristic by one. Consequently $\chi(N_q) = 2 - q$, where q is the number of holes cut out.

A closed nonorientable surface which is interesting and important for later examples is the projective plane $P^2 = N_1$. In geometry the projective plane is defined using a system of homogeneous coordinates. A point x in the projective plane P^2 has coordinates (x_0, x_1, x_2), where x_0, x_1, x_2 are real numbers, at least one of which is not equal to zero. The term "homogeneous" means that two points $x = (x_0, x_1, x_2)$ and $x' = (x'_0, x'_1, x'_2)$ represent the same point of the projective plane when they both lie on a straight line passing through the origin.

Thus the projective plane is the set of unoriented straight lines in three-dimensional Euclidean space. If one draws a sphere about the origin in three-dimensional space, every straight line intersects it at two points, which thus must be considered equivalent. One can prove that the mapping so created from the sphere onto the projective

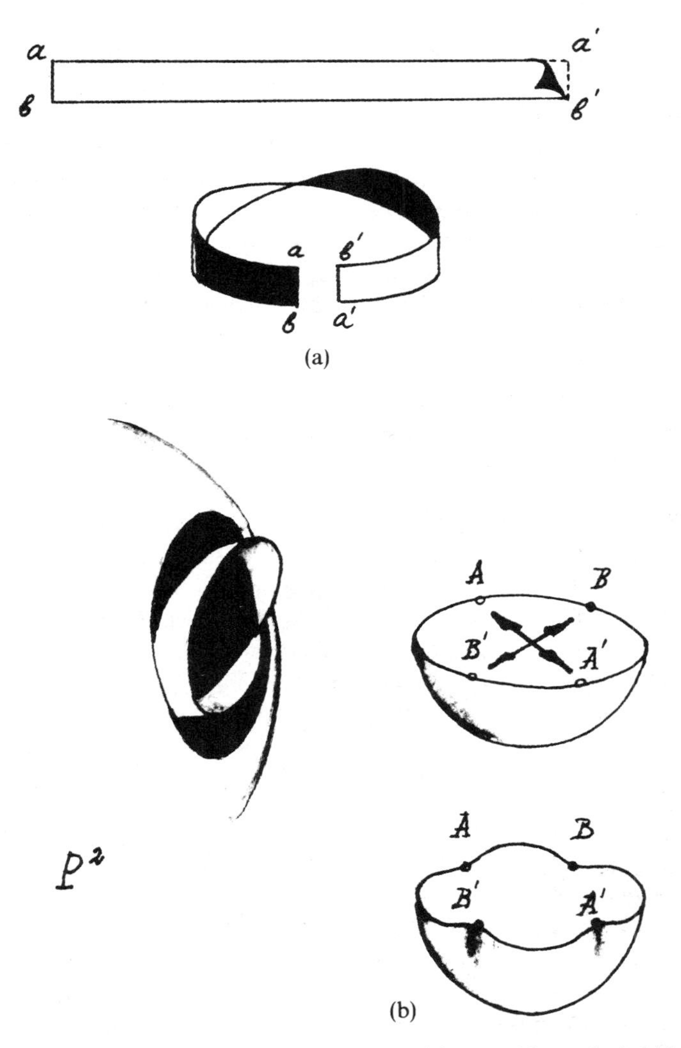

Figure 9.3: Construction of nonorientable surfaces. (a) A Möbius strip $(abb'a')$. Sides ab and $a'b'$ are identified after rotating one of the sides (for example, $a'b'$) by 180°. (b) Construction of the closed nonorientable surface P^2. The edge of the Möbius strip can be glued to one boundary of a "hole" in the sphere. Such pasting together cannot really be done in three-dimensional space unless the surfaces are allowed to intersect themselves.

plane is continuous. In consequence a projective plane is topologically equivalent to a sphere with the antipodal points identified (two such points on a sphere define one point of a projective plane). This definition will be convenient in the subsequent exposition.

By cutting a sphere along the equator, we obtain two hemispheres. Lets consider one of them. In order to obtain a projective plane from this hemisphere we have only to identify diametrically opposite points on the boundary circle—the edge of the hemisphere. Such an identification is equivalent to pasting on a Möbius strip. (Remember the definition of a Möbius strip.) Consider the Euler characteristic of the projective plane. The projective plane is equivalent to a hemisphere with a Möbius strip pasted on. The transformation from a sphere to a hemisphere is equivalent to cutting a hole, that is, to reducing the Euler characteristic by one, while pasting on a Möbius strip does not change it. Thus the Euler characteristic of the projective plane is equal to one.

A still better-known example of a closed one-sided surface is the Klein bottle (Fig. 9.4). Like the Möbius strip, it also is obtained by a special identification of the sides of a rectangle. Consider the rectangle $ABCD$. From this rectangle we paste together a cylinder by identifying sides AB and CD. The cylinder has two edges, circles l_1 and l_2. If one simply pastes these circles together, the result is an orientable surface, a torus. The Klein bottle is formed by identifying the circles l_1 and l_2 after reflecting l_1 about the diameter ab. The Euler characteristic of the Klein bottle is equal to zero.

One can prove (with difficulty) that any closed two-dimensional surface is homeomorphic to one of those enumerated above. Thus, two-dimensional surfaces are characterized by two parameters: genus and orientability.

It would be very strange if mathematicians confined themselves to the study of two-dimensional manifolds and made no attempt to classify multidimensional spaces. The task of classifying multidimensional spaces is incomparably more complex. The classification of even three-dimensional manifolds is a difficult unsolved problem, and the problem of classifying four-dimensional spaces is in a certain sense unsolvable. The Russian mathematician Andrei A. Markov, Jr. (1903–1979) proved the impossibility of constructing an algorithm that would make it possible to compute whether or not two given four-dimensional manifolds are homeomorphic. In the early 1980's the American mathematicians William P. Thurston and Michael H. Freedman and the British mathematician Simon Donaldson have made great progress in the classification of both three- and four-dimensional manifolds. The results of Donaldson, who has constructed various smooth structures on simply connected 4-dimensional manifolds, use constructions from the theory of gauge fields. Some idea of these remarkable discoveries will be given below in the chapter on "Topological Particles."[1]

The structure of manifolds of higher dimensions has special interest for quantum physics. Examples of such manifolds arise in quantum field theory and quantum gravity, where the basic objects of study are 4-dimensional space-time, gauge fields, and multidimensional symmetry groups.

In higher dimensions the classification problem arises only for manifolds endowed with additional structures. Topology now has powerful methods of studying specific

[1]For expository accounts of their work, see: William P. Thurston, "Three dimensional manifolds, Kleinian groups, and hyperbolic geometry," *Bull. Amer. Math. Soc.* (NS), **6** (1982), 357–381, and Michael H. Freedman, "There is no room to spare in four-dimensional space," *Amer. Math. Soc. Notices* **31** (1984), 3–6.

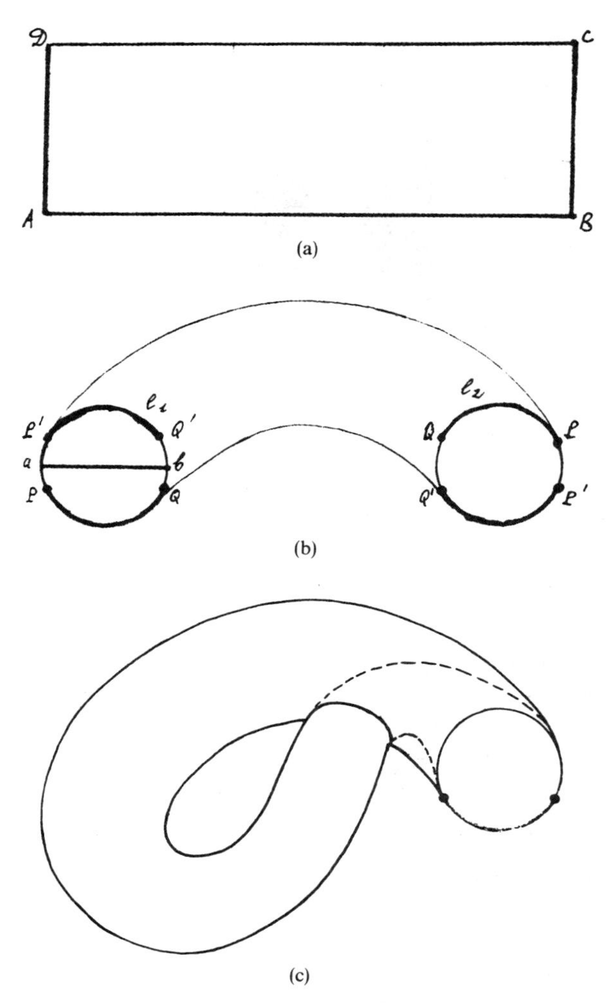

Figure 9.4: The Klein bottle—a closed nonorientable surface of genus 0. (a) The initial rectangle $ABCD$. (b) The cylinder obtained by gluing sides AB and CD together. (c) The identification of the circles. (d) The Klein bottle. The boundaries l_1 and l_2 are identified after a reflection with respect to the diameter ab.

manifolds. We mention here only two classes of invariants, which in a number of cases yield a simple solution to the problem of determining the topological structure of a given manifold. These two approaches, which are characteristic of topology, call to mind an analogous situation in physics. Physical bodies can be studied in two ways: through their internal structure or through their interaction with known objects. A typical example is the measurement of an electric field by introducing a standard charge into it. In topology, homology theory corresponds to the former,

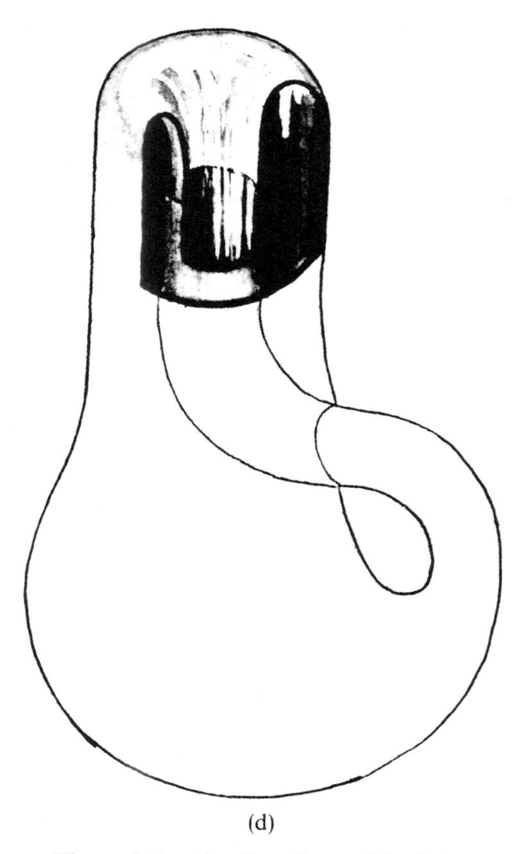

(d)

Figure 9.5: Continuation of Fig. 9.4.

while homotopy theory corresponds to the latter. The connection between the two theories is quite subtle.

Homology theory studies properties of manifolds by decomposing them into simpler parts. The structure of these parts can be investigated easily by introducing algebraic characteristics associated with these decompositions. The main difficulty lies in proving that the corresponding characteristics of the decomposition, in fact, do not depend on the particular choice of the decomposition but are rather a topological invariant of the manifold itself.

The Euler characteristic already discussed can serve as an example of a topological invariant of a homological type. Let us look at two regular polyhedra: the tetrahedron T and the cube C. We shall compute the Euler characteristic ($\chi = V - E + F$) of the tetrahedron T and the cube C. For the tetrahedron, $\chi(T) = 4 - 6 + 4 = 2$, while for the cube, $\chi(C) = 8 - 12 + 6 = 2$. Let us describe a sphere around each of the polyhedra and project the polyhedra onto the surface of the sphere. As a result

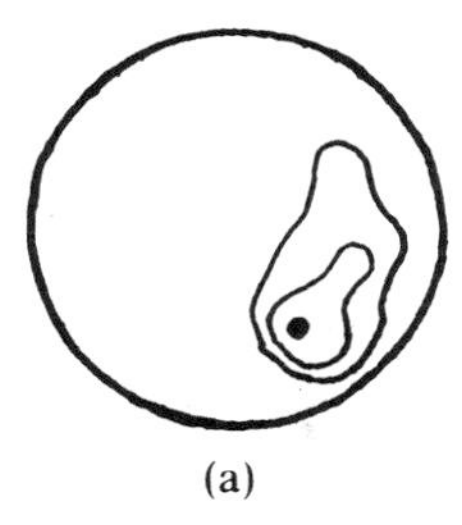

(a)

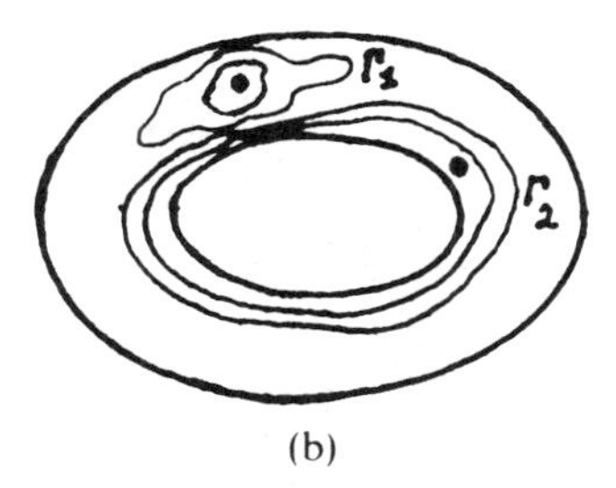

(b)

Figure 9.6: Contours on a disk and an annulus. (a) A disk. Any closed path can be contracted to a point. (b) An annulus. Path Γ_1 can be contracted to a point, while path Γ_2 cannot.

we obtain two grids on the sphere. These grids produce different decompositions of the sphere, but, as is obvious, they have one and the same Euler characteristic. By definition this number is the Euler characteristic of the sphere.

If we know in advance that the necessary characteristic is a topological invariant, we can use the simplest decomposition of the manifold to compute it. The computation of the invariant for this decomposition thereby defines it for the original manifold.

Homotopy theory supplies us with invariants of the other type. Let us look at two examples of homotopy invariants. Take two figures, a disk and an annulus (Fig. 9.6). Draw a closed curve Γ in each of them. Now try to contract the curve Γ to a point by a continuous deformation. It is obvious that this is possible only in the disk. In the annulus the path Γ_2 can be contracted to a circle enclosing the hole, but there is a closed curve Γ_1 that can be contracted to a point. It is easy to show that all closed paths in an annulus contract either to a point or to a path which goes around the boundary circle, possibly several times. In the disk all closed paths can be contracted to a point. A space in which all closed paths can be contracted to a point is called *simply connected*. Of the two-dimensional orientable surfaces, only a sphere is simply connected. For example, on a torus there is an infinite set of closed paths that cannot be contracted to a point. Moreover, there exist closed paths not contracting to a point which cannot be continuously deformed into each other (Fig. 9.7). Such paths are called nonhomotopic. The set of classes of closed paths, nonhomotopic to one another—a topological invariant of a manifold—is called the *fundamental group* of the manifold. This group is also called the first homotopy group π_1, because it is the first in a whole sequence of homotopy invariants, the homotopy groups π_n.

In what follows we shall frequently have to deal with various groups. Let us recall the definition of a group. Consider a set G of elements g for which an algebraic operation " $\times$ " called multiplication, is defined. Specifically, for each two elements $g_1, g_2 \in G$, the element $g_3 = g_1 \times g_2$ is a well-defined element of G. The set G is called a group if it fulfills the following conditions:

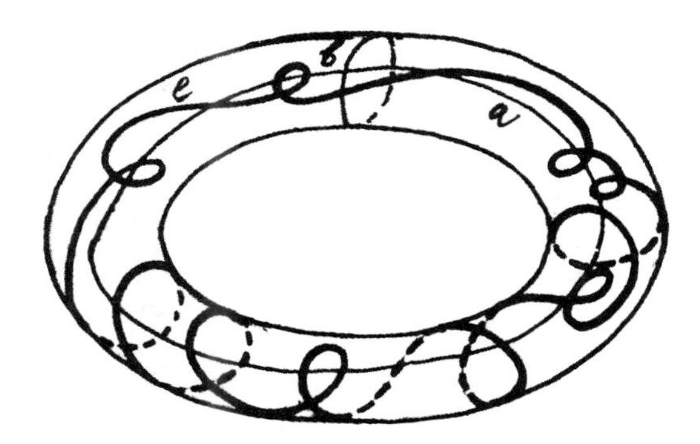

Figure 9.7: Nonhomotopic paths on a torus. Path a is not homotopic to path b. An arbitrary closed path c on the torus is homotopic to a path passing n times along the parallel a and m times along the meridian b.

1. for any three elements g_1, g_2, g_3, the identity $g_1 \times (g_2 \times g_3) = (g_1 \times g_2) \times g_3$ holds (associativity of multiplication);

2. G contains an element e (the identity of the group) such that for any $g \in G$, $e \times g = g \times e = g$;

3. for any element $g \in G$ there exists an element g' such that $g \times g' = g' \times g = e$. The element g' is called the *inverse* of g and is denoted g^{-1}.

The definition of a group does not require that $g_1 \times g_2 = g_2 \times g_1$ for all $g_1, g_2 \in G$. If this requirement is met, the group is called *commutative* or *Abelian*, in honor of the famous Norwegian mathematician N. H. Abel.

The set of symmetries of an equilateral triangle, the group T_3, serves as a useful illustrative example of a group. It consists of six elements: three rotations g_1, g_2, g_3 by angles $2\pi/3$, $4\pi/3$, and 2π respectively, and elements g_4, g_5, g_6 given by $g_{i+3} = sg_i$, $i = 1, 2, 3$, obtained by combining a rotation with a reflection s about an axis of symmetry of the triangle ($s \times s$ is the identity of the group). The set of pure rotations (without reflections) forms a subgroup R_3 of the group T_3. It follows from this that the product of two rotations will be another rotation. The subgroup R_3 is a commutative group, but the full group T_3 is noncommutative, as the reader can easily see.

The fundamental group was discovered by Poincaré and is sometimes called the Poincaré group by mathematicians. This name would be inconvenient to use in physics literature, however, where the Poincaré group is the accepted name for the full group of transformations of space-time.

It is convenient to put the definition of the fundamental group in a form which allows us to construct a multidimensional generalization, the n-dimensional homotopy group π_n. A closed path Γ on a manifold M can be represented as the image of a fixed circle S^1 (the "test body"). Then the fundamental group $\pi_1(M)$ is the set

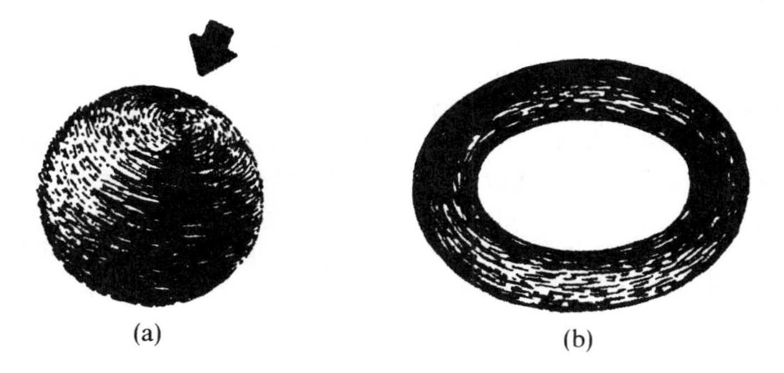

Figure 9.8: Vector fields on a sphere and a torus. (a) On the sphere. (b) On the torus.

of mappings of the circle S^1 into the manifold M which are not homotopic to each other. If we now replace the circle in the definition of the fundamental group by the n-dimensional sphere, we obtain the definition of the n-dimensional homotopy group $\pi_n(M)$.

Between the homology and homotopy characteristics, there exist various links. For example, a simply connected manifold was defined above by the requirement that the fundamental group be equal to zero. On the other hand, Riemann's initial definition of connectivity using a system of cuts had the character of homology. Let us give a homological definition of simple connectivity. A surface is called *simply connected* if any closed path separates it into disjoint parts. From this definition and the classification of two-dimensional surfaces, it also follows that the sphere is the only simply connected orientable two-manifold. The coincidence of two such different definitions is not accidental but rather the result of a fundamental topological theorem proved by the Dutch scholar Witold Hurewicz (1904–1956).

The topological type of a manifold imposes considerable limitations on the behavior of vector, tensor, and other fields defined on it. How, for example, can we construct a continuous nonzero field of tangent vectors on a closed two-dimensional surface M^2? In other words, is it possible to comb a hedgehog (Fig. 9.8). Let us impose a minor constraint on nature and suppose that the hedgehog is a two-dimensional orientable surface. (In the nonorientable case yet another hedgehog will appear in the guise of the Klein bottle.) After combing, not a single spine of the hedgehog is to jut out beyond the normal surface; more precisely, if from each point there grows a spine (a nonzero vector, not necessarily tangential to M^2), and the direction of the spines is a continuous function of the coordinates of the point x_0 lying on the surface, then, after combing, not a single spine is to be directed perpendicular to the surface M^2. The answer depends on the topology of the surface. It turns out that one can comb only a "torus" hedgehog.

The mathematical formulation of this fact consists of confirming that a tangential vector field degenerates, that is, it must be equal to zero for at least one point of any surface except the torus. In such a case the vector field is said to have critical points.

Poincaré proved a remarkable theorem connecting the number of critical points of a vector field on a surface with the Euler characteristic of the surface. A vector field without critical points exists on a surface if and only if the Euler characteristic of the surface is equal to zero. The generalization of this problem, that is, the construction of vector and tensor fields on manifolds of higher dimension with a given type of critical point, constitutes one of the major branches of topology—the theory of characteristic classes.

We shall now illustrate the connection of topological properties of spaces with fundamental physical phenomena by a simple, well-known example.

Chapter 10

The Connectivity of a Manifold and Quantization of Magnetic Flux

If a ring in a superconducting state is placed in a magnetic field and then the field is turned off, a superconducting current will begin to flow in the ring. In striking fashion, it turns out that the magnitude of magnetic flux is quantized, that is, it takes on only values from the discrete set of numbers $cnh/2e$, $n = 0, 1, 2, \ldots$, where h is Planck's constant, e is the charge of an electron, and c is the speed of light. In the case of continuous superconductivity, the flux is equal to zero. This result follows from the macroscopic theory of superconductivity, supplemented by the concept of the Cooper pairing of electrons. This result, discovered by the American physicist Leon Cooper, is the basis of the contemporary microscopic theory of superconductivity, established by J. Bardeen, L. Cooper, and J. Schrieffer in 1957 (the BCS theory, for which the three received the 1972 Nobel Prize).

Briefly, the phenomenon can be described as follows: Electrons in a superconducting state attract one another, forming connected pairs—Cooper molecules. These molecules should not be pictured as united point particles, however. Two electrons forming a pair can diverge to distances greater than the average distance between pairs. Nonetheless, one can talk about the wave function of the pair ψ and the spin of the pair, which is equal to zero. This is the theoretical difference between superconductivity and the usual flow of free electrons—particles with a spin of $1/2$. As is well known, fermions—particles with a spin of $1/2$—obey the Fermi–Dirac statistics, that is, only one particle can be found in each state. Bosons, which are particles with integral spin (especially zero), are subject to the Bose–Einstein statistics, that is, any number can be found in each state. The formation of a large number of bosons in the lowest energy state is called the *Bose condensate*. The existence of the condensate at a given temperature T_c (the temperature of superconductivity) leads to a superconducting current.

The concept of the Bose condensate makes it possible to consider a series of effects, among them the quantization of magnetic flux using macroscopic quantities like the wave function of superconducting pairs ψ. Formally the function ψ satisfies

$$-\frac{\hbar}{i}\frac{\partial}{\partial t}\psi = \frac{1}{2m}\Big(\frac{\hbar}{i}\nabla - \frac{qA}{c}\Big)^2 + q\phi\psi, \tag{10.1}$$

the usual Schrödinger equation for a nonrelativistic particle with zero spin moving in an electromagnetic field. But the charge q means the charge of a pair, $q = 2e$, where $\hbar = h/2\pi$, A is the vector potential, and ϕ is the electric potential of the electromagnetic field. The current density j is related to the charge density $\rho = |\psi|^2$ by the equation

$$j = \frac{\hbar}{m}\Big(\nabla\theta - \frac{1}{\hbar c}A\Big)\rho. \tag{10.2}$$

Here $\theta = \arg\psi$ is the phase of the wave function $\psi = |\psi|e^{i\theta}$.

It is well known that the current density j inside a body found in a superconducting state is equal to zero. Formula (10.2) implies the relationship between the phase and the vector potential A,

$$c\hbar\nabla\theta = qA.$$

If the conductor is not continuous, then the contour integral on any closed path enclosing a hole equals

$$c\hbar\oint\nabla\theta = q\oint A\,dS = q\int \operatorname{curl} A\,d\sigma = q\Phi.$$

The current Φ in a ring superconductor is equal to the value of the change of phase of the wave function ψ as it traverses a closed contour. It is natural to require that the wave function have unit length at every point. Then as it travels around the closed contour, the phase θ can change only by $2\pi n$, where n is an arbitrary integer, and the magnetic flux Φ across the ring superconductor will be a multiple of the number $c\hbar/q$.

This effect was predicted in 1948 by the American physicist F. London (1900–1954), even before the creation of the microscopic theory of superconductivity. The magnitude of a minimal flux was predicted to be twice what it proved to be experimentally. London had considered the magnitude of q equal to the charge of the electron e.

Let us now consider this effect from the point of view of topology. The integral of the gradient of the phase along an arbitrary path in a disk is equal to zero, because one can deform this path to a point, but in a ring such a path is equivalent to a circle enclosing the hole and, as was shown above, the integral along it is equal to $q\Phi$.

In view of the fact that a disk and a ring have different connectivity, London's effect is formulated as follows. Quantum flux exists in a multiply connected conductor and is absent in a simply connected conductor. Quantization of flux thus is connected with the topology of a superconductor. The example just analyzed is only the first of a whole series of physical effects with topological components. Some of these will be discussed in subsequent chapters.

Chapter 11

Systems with Spontaneous Symmetry Breaking

MANY physical systems, seemingly quite different from one another, turn out to be susceptible to investigation by certain topological methods. The phenomenon that unifies such physically different systems as liquid crystals, magnetism, and superfluid helium, is called *spontaneous symmetry breaking*. It underlies many contemporary concepts in the theory of elementary particles, the theory of phase transitions, and a number of problems of cosmology.

The essence of this phenomenon is revealed by a particular example taken from the theory of magnetism. Given a low enough temperature, a wide variety of crystals become magnetized in the absence of an external magnetic field. This phenomenon is called *ferromagnetism* and is explained by the existence of a special exchange interaction among the atoms of crystal lattices. The magnetization thus created is called spontaneous because it is formed without the application of an external field and is characterized by the magnetization vector M—the magnetic moment of a ferromagnet.

We will not delve more deeply into the labyrinth of the theory of the ferromagnet but will consider the classical model—the isotropic Heisenberg ferromagnet—as a spontaneous symmetry breaking effect. Consider a crystal lattice, at the vertices of which there are localized particles with half-integer spin, for example, electrons. For the sake of simplicity, we shall consider the lattice to be two-dimensional, although our conclusions are valid also for a three-dimensional lattice (Fig. 11.1). The interaction between electrons located in neighboring points of the lattice is defined by spin vectors $\mathbf{S}(x)$. For the sake of definiteness we take the Hamiltonian (energy operator) H to be

$$H = \sum_{x,x'} I(x - x')\mathbf{S}(x)\mathbf{S}(x'). \tag{11.1}$$

The function $I(x) = 0$ if $x \neq \mathbf{a}$, where $\mathbf{a}$ is a base of the lattice $I(\mathbf{a}) = \lambda$ ($\lambda < 0$). The quantity $I(x)$ is called the *exchange integral*, and the interaction itself is called a nearest-neighbor interaction, because the contribution in (11.1) involves only

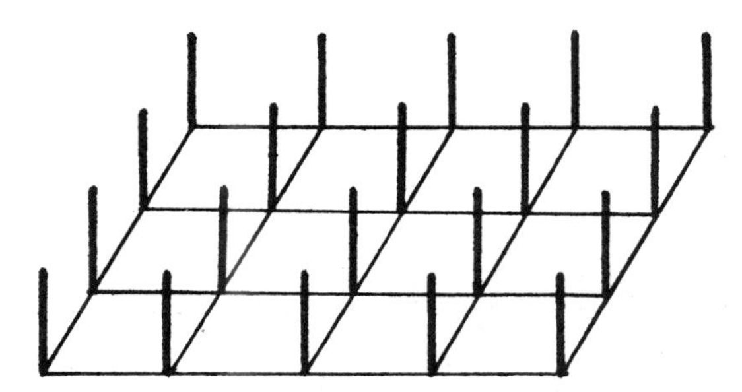

Figure 11.1: The isotropic Heisenberg ferromagnet on a two-dimensional lattice. The ground state of the system $\{M^2 = \max\}$. (All spins **S** look "upwards").

interactions between two adjacent points. The magnetization vector M has the form

$$M = \sum_x \mathbf{S}(x). \tag{11.2}$$

One can show that the Hamiltonian (11.1) does not vary with rotation of the magnetization vector M; that is, the energy does not depend on its direction. The state with the least energy—the ground state of the system—corresponds to the greatest value $\langle M^2 \rangle$. Remember that $I < 0$. In this state all spins S are oriented identically along a certain fixed axis $\mathbf{n}$ and the projection of the magnetization vector M in the direction of the orientation of the spins $\mathbf{n}$ has a definite value. From this it obviously follows that the ground state is not invariant relative to the full group of rotations of the vector M. The symmetry group must preserve the fixed direction of the vector M. In the given case, it corresponds to the group of rotations of a circle.

Now, at last, we can give the definition of spontaneous symmetry breaking. Systems in which the symmetry of the ground state does not correspond to the symmetry of the Hamiltonian are called systems with spontaneously broken symmetry. This is the accepted name for such systems, but it would be more correct to call them systems with *hidden* symmetry. In essence, the symmetry of the Hamiltonian is not broken, only hidden. In the ground state it is impossible to uncover the higher symmetry of the system.

Examples of similar symmetry-breaking are encountered in various problems of physics. It is known, for example, that nuclear forces are invariant relative to rotations; at the same time, the ground state of a nucleus with nonzero spin is not invariant relative to the group of rotations.

The effect of spontaneous symmetry breaking is one of the mechanisms that explain a broad range of phenomena—phase transitions in matter.

It is known that one and the same substance can be found in different states or phases, depending on external conditions (temperature, pressure, etc.). The transition from one phase to the other is called a phase transition. Phase transitions occur in a wide variety of materials. As an example, the transition of a metal from the normal state to the superconducting state occurs at very low temperatures; in mercury, for example, the first substance in which the phenomenon of superconductivity was observed, electric resistance drops to zero at a temperature $\sim 4\,\mathrm{K}$. On the other hand, in recent years views on the nature of superconductivity have greatly altered as a result of the striking discovery of a class of superconductors in the Ba–La–Cu–O family with phase transition temperature $T_c \sim 30\,\mathrm{K}$ in 1986 by the Swiss physicists J. Bednorz and A.K. Müller, winners of the 1987 Nobel Prize. In later experiments Curie temperatures as high as $T_c \sim 90\,\mathrm{K}$ have been attained. As another remarkable example helium ^{4}He becomes superfluid at temperatures around (2 K). In 1972 a phase transition into the superfluid state was discovered in another isotope of helium, ^{3}He. It occurs at a fantastically low temperature—$2.6 \times 10^{-3}\,\mathrm{K}$.

Underlying the contemporary theory of phase transitions is the study of the symmetry of systems in various phases. More precisely, it is known that phase transitions of first and second kind occur. The first kind includes the classical examples: solids dissolve or are formed, liquids evaporate, gases condense. The process of melting ice at great pressure is such a transition. Transitions of the second type include the formation of superfluid helium from normal helium and the transition of metal into a superconducting state.

By their nature phase transitions of the first and second type differ substantially. One can explain the difference by referring to the classical 1937 work of Lev Landau (1908–1968), which is the foundation of the contemporary theory of phase transitions. Landau regarded phase transitions as changes of the symmetry of matter. For a quantitative description of a phase transition, he introduced a degree of ordering, the *order parameter*. The order parameter is defined differently for specific systems but possesses an important general property: it is equal to zero in a "disordered" phase and is nonzero in an "ordered" phase.

Phase transitions of the first and second kind are distinguished by the behavior of the order parameter. In transitions of the first kind the order parameter changes by jumps, in the second kind it changes continuously.

Let us examine the phase transition of ice to water, the process of melting, from this perspective. If one chooses the order parameter η to be the ratio of the number of molecules at the nodes of a crystal lattice to the total number of molecules, then in the "ordered" phase (ice) η is nonzero, but in the "disordered" phase (water) $\eta = 0$. In this transition, η changes by a jump, and consequently, the melting process is a phase transition of the first kind.

We shall now give a precise meaning to the terms "ordered" and "disordered" phases. Consider the symmetry of two states, ice and water. It is natural to regard the symmetry group of the crystal lattice Γ as the symmetry group of ice. The group Γ is discrete. Its elements are translations of the lattice, spatial rotations and reflections. The symmetry group of water G is considerably larger; by this is meant the group of all transformations that preserve the hydrodynamic equations of an incompressible fluid. The group G is a continuous infinite-dimensional group.

In our example the symmetry group G of the less ordered phase, water, contains as a subgroup the symmetry group of ice Γ. A similar situation obtains in phase transitions of the second kind. In phase transitions of the first kind, it may happen that the symmetries of the two phases are not connected by any sort of relationship.

To take another example, let us consider the now-familiar Heisenberg ferromagnet (a transition of the second kind) from the perspective of phase transitions. In the ground state of the ferromagnet ($T = 0$) all spins are directed identically. We have defined this direction by the vector $\mathbf{n}$ which coincides with the direction of the magnetization vector M. Under heating ($T > 0$) the correlation between spins S_i is weakened; their orientation becomes chaotic, and correspondingly the mean value $\langle M^2 \rangle$ decreases. At a certain temperature T_c, the *Curie point*, the value of $\langle M^2 \rangle$ reaches zero. A ferromagnet loses its magnetic properties and converts to paramagnetic phases. One can consider the magnetization vector M as the order parameter defining the phase transition in this system.

We must emphasize again the connection of symmetry with the properties of the ordered and disordered phases. In the ordered, or ferromagnetic phase, the system has the symmetry of the ground state. This symmetry group coincides with the group of rotations, leaving as invariant the direction of magnetization along which spins were lined up. Transition to the paramagnetic state results in a system with a larger symmetry group, the group of all rotations of three-dimensional space. This is because in the paramagnetic phase there is no preferred direction. In this phase the full symmetry group of the Hamiltonian coincides with the symmetry of the ground state, or (as they say in physics), symmetry has been restored.

In this example we see all the peculiarities of the general outline: the symmetry group corresponding to the ordered phase is smaller than that of the disordered phase. In this case the symmetry group of the disordered phase contains the symmetry group of the ordered phase as a subgroup.

Landau's macroscopic theory of phase transitions has made it possible to describe a very broad range of phenomena. Nonetheless the difficult problems involved with a detailed description of phase transition near critical points remain unsolved. In recent years substantial progress has been made in this area, associated with the application of the ideas of field theory, but the problem of constructing a microscopic theory of phase transitions is still far from being definitively solved.

The range of problems for which a solution is needed, if only on a macroscopic level, has expanded remarkably. It is precisely here that topology ought to aid physicists. As an example we shall show how the problem of the existence of line (thread-like) and point singularities in liquid crystals can be solved using topological methods. This example was not chosen haphazardly. First, the study of liquid crystals is an appealing problem, important both in a theoretical and an applied sense. Second, the peculiarities of applying topological methods in the physics of liquid crystals are typical of many systems with spontaneous symmetry breaking.

Chapter 12

Topology and Liquid Crystals

LIQUID crystals are structures in an intermediate state between a liquid and a rigid body. There is a certain ambiguity in the very term. True liquid crystals have both the properties of a crystal (orderliness of structure) and the properties of a liquid (fluidity).

The atoms in a crystal are located at the nodes of a regular three-dimensional lattice. This property was always assumed by crystallographers, but only after the classical experiments on the diffraction of X-rays by Max von Laue (1879–1960) in 1912 was there the possibility of "viewing" the lattice. The British scientists William Henry Bragg (1862–1942) and William Lawrence Bragg (1890–1971), father and son, developed a method for determining the structure of a crystal lattice by measuring the intensity of the scattering of X-rays from it. In an X-ray diffraction image, repeating dots (Bragg images) appear at certain distances which are characteristic of the lattice. In liquid crystals, a certain orderliness also arises, but, in contrast to solid crystals, the periodic structure is usually observed only in one or two dimensions. Both dimensions occur in nature.

A large class of liquid crystals called *smectics* possess spatial orderliness only in one dimension. The mechanical properties of smectic liquid crystals resemble the properties of soap. (The term *smectic* was proposed by the French crystallographer Georges Friedel (1865–1933) and comes from the Greek $\sigma\mu\acute{\eta}\gamma\mu\alpha$, meaning *soap*.) In their structure smectics are like a pile of leaves whose layers are an equal distance apart. Each layer is a two-dimensional liquid. This class of liquid crystals, despite all its remarkable properties, is somewhat peripheral to the topological applications we wish to study. Let us leave smectics for a while and look more closely at other forms of liquid crystals.

A large number of liquid crystals with properties completely different from smectics are formed from organic molecules; they are called *nematic* liquid crystals. One of the commonest examples of a nematic liquid crystal is *n*-azoksianizol.

This crystal is a rigid rod with length about 20 Å and thickness about 5 Å. Given a sufficiently high temperature, (120° C) and atmospheric pressure, molecules of *n*-azoksianizol are found in the nematic liquid-crystal phase.

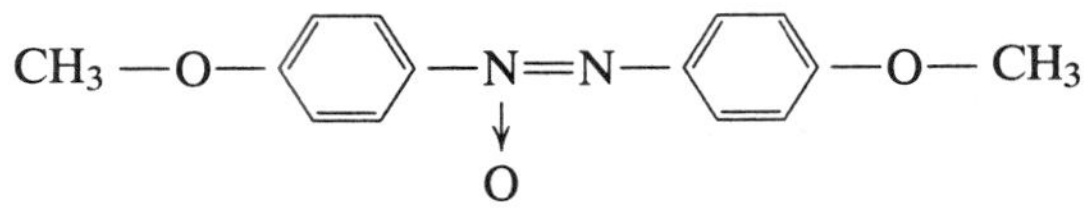

$$CH_3 - O - C_6H_4 - \underset{\underset{O}{\downarrow}}{N} = N - C_6H_4 - O - CH_3$$

Figure 12.1: The nematic n-azoksianizol.

The term *nematic* (from the Greek $\nu\tilde{\eta}\mu\alpha$, meaning *thread*) was also invented by Friedel, who observed the threadlike defects characteristic of such crystals. In their structure nematics are closest to liquids, but they differ sharply from them in their optical properties. The most natural way to think about nematics is as a collection of sticks or distended molecules strewn about in space. At the same time, the centers of gravity of the molecules are located completely arbitrarily, and in this they are similar to an ordinary liquid, where there is no particular relationship between the centers of molecules. However, there exists a certain order in the orientation of the molecules. They are lined up parallel to a certain axis characterized by the unit vector $\mathbf{n}$. The direction of the vector $\mathbf{n}$ coincides with the direction of the optical axis of the medium (that is, the major axis of the molecule). In contrast to an ordinary liquid, a nematic behaves like a doubly refracting medium. The difference between the indices of refraction for polarization along the optical axis and in the direction perpendicular to it is very large.

Nematic phases are found only in liquid crystals that are optically pure—that is, all the molecules in the solution rotate polarized light in the crystal in the same direction under the same conditions. The instant a stereoisomer (mirror image) molecule is introduced, a spiral distortion appears in the liquid crystal structure. This phenomenon was first observed in cholesteric ether; for that reason this phase is called *cholesteric*.

To obtain quantitative information about the behavior of such systems in different phases, we introduce an order parameter, by means of which one can construct a theory of transitions from a more ordered to a less ordered phase.

Phase transitions in a nematic crystal are similar in many ways to the transition from a ferromagnetic to a paramagnetic state. Nonetheless, despite many analogies between them, there is a substantial difference in the nature of their phase transitions. The transition from the nematic phase to the isotropic is a transition of first kind, although the size of the jump of the order parameter is quite small.

The transition from the ordered-nematic phase to the disordered-isotropic occurs at a higher temperature and is connected with a change in the symmetry of the system. The symmetry of the nematic phase is lower than the symmetry of the isotropic state—that is, the symmetry group of the isotropic phase contains the symmetry group of the nematic phase as a subgroup.

Many properties of nematics can be investigated in the framework of common regularities inherent in systems with spontaneous symmetry breaking. The first problem that arises here is to determine an order parameter. The order parameter will be a vector $\mathbf{n}$ that characterizes the direction of the molecules in the nematic phase. As with the Heisenberg ferromagnet, in the nematic phase the full symmetry occurs with respect to rotation about the vector **n**. In the isotropic phase, as we recall, molecules are oriented completely arbitrarily, consequently, there is no preferred direction of spatial orientation. The enlargement of the symmetry group of the isotropic liquid is obviously due to this circumstance.

Let us recall the analogous situation in a ferromagnet. The order parameter in nematics has an additional property which, as we shall see subsequently, is of fundamental significance. The states of a nematic system defined by the vectors **n** and $-\mathbf{n}$ are indistinguishable. For example, if an individual molecule has a permanent dipole moment, then the number of dipoles directed "downwards" is exactly equal to the number of dipoles directed "upward." We now have all the necessary construction material to describe a remarkable phenomenon, the existence of two types of defects in nematic systems: point and line defects.

In sufficiently thick specimens of a nematic, line defects are visible as a system of dark flexible threads having a very odd configuration. Some threads move by bending and forming a self-intersection while others are less mobile, being fastened by their ends to the walls of the container. The first investigators to observe this thread formation were the French physicists François Grandjean (1882–1975) and G. Friedel, who believed that the phenomenon of threads is associated with the structure of a nematic and not with external causes, but it was many years before these beliefs were verified.

Let us now turn from experiment to theory and try to explain the emergence of threadlike singularities in a nematic. This can be done by studying the topology of the space of the order parameter. The order parameter, the vector **n**, completely determines all the states of a nematic. A vector **n** is given at each point of the nematic and a priori is oriented completely arbitrarily. The only other restriction is that the states defined by the directions **n** and $-\mathbf{n}$ are indistinguishable. A vector **n** satisfying this last condition is called a *director*.

Do there exist lines and points in a nematic where the director **n** has discontinuities? That is, do there exist threadlike and point singularities? We shall see how topology helps to gain an understanding of line (threadlike) singularities. Suppose such a thread L exists. We encircle it with a closed contour γ and map γ into the domain of values of the order parameter. The domain of values of the order parameter is the domain of variation of the director **n**. This domain is a sphere with antipodal points identified. We already know that this surface is homeomorphic to the projective plane P^2. Corresponding to every point x of the contour the director **n** determines a point x_1 in the projective plane. Given this, the contour γ is mapped into a closed curve γ_1. Let us contract the contour toward the line L (Fig. 12.2). The image of the contour γ, the curve γ_1, will contract toward some curve γ_2.

There are two possibilities: Either the curve γ_1 can be contracted to a point, or it cannot. If the curve γ_2, and therefore also γ_1, can be contracted to a point $a \in P^2$, then the vector field $\mathbf{n}(x)$ defined on the curve γ, can be deformed continuously to the constant field $\mathbf{n}(x) \equiv a$. In this case it is clear that the line L is a removable line singularity, because contracting the contour γ to the point x_0 on L causes the continuous field $\mathbf{n}(x)$ to contract to a point lying on the line. In the second case, the situation is completely different. One cannot get rid of singular lines by continuous transformations of the field.

The reasoning above provides a simple criterion for the existence of line singularities or, as crystallographers say, disclination. There are no line singularities if all closed curves in the space of an order parameter can be contracted to a point. Thus, in the case of a nematic, the question of the existence of disclinations reduces to a purely topological problem: *Do there exist closed paths not contractible to a point on a projective plane?* In contrast to the usual sphere, such paths do exist. Any path (we will denote it by γ_3) joining two antipodal points of the sphere defines a closed path on P^2 which satisfies this requirement. One can prove that all closed paths on a projective plane contract either to a point or to the path γ_3.

In physics removable disclinations are spoken of as unstable structures. Distortions around such lines can always be continuously transformed into smooth structures, but this is not possible for disclinations that correspond to paths not contractible to a point.

The distinctions just noted among disclinations can be clearly observed in an experiment. Under a microscope threadlike singularities are studied in thin films of nematic liquid. In polarized light, as was shown in the experiments of C. Williams and Y. Bouligand, unstable disclinations (corresponding to contractible paths) appear as wide dark bands, but stable disclinations (uncontractible paths) are narrow bands.

To study critical points in a nematic one proceeds as follows. A point x where the vector $\mathbf{n}(x)$ is not defined is surrounded by a sphere. Then the problem, as in the case of threadlike singularities, is reduced to a purely topological one: *How many homotopically distinct contraction mappings of sphere onto projective space exist?* It turns out there exists an infinite set of such mappings, characterized by integers.

Recalling the already familiar definition of homotopy groups, we see that the problem of classifying linear and critical points comes down to determining the first and second homotopy groups of the space of an order parameter. In topology, a technique for computing these groups was developed long ago.

With such methods one can investigate critical points in cholesteric liquid crystals, superfluid helium, ferromagnets, and the like. Of course, we did this in somewhat simplified form. In order to be convinced definitively of the existence of critical points, one must investigate the extent to which similar configurations are stable and minimize energy. If the topology of the system implies that there should be no critical points, however, none can exist, at least not for long.

Liquid crystals are an ideal object for applying various topological methods. A more sophisticated analysis makes it possible to study the processes of adhesion, dissociation and collapse of threads in nematics and cholesterics.

Interesting applications of topology occur in the study of critical points associated with distinct boundary effects. Do critical points occur (for example, vortices) on the surface of nematics or smectics? One type of critical point in a nematic has already been examined, but point defects can also occur on the surface of a nematic, either as isolated points on a surface or as the ends of disclinations inside a body.

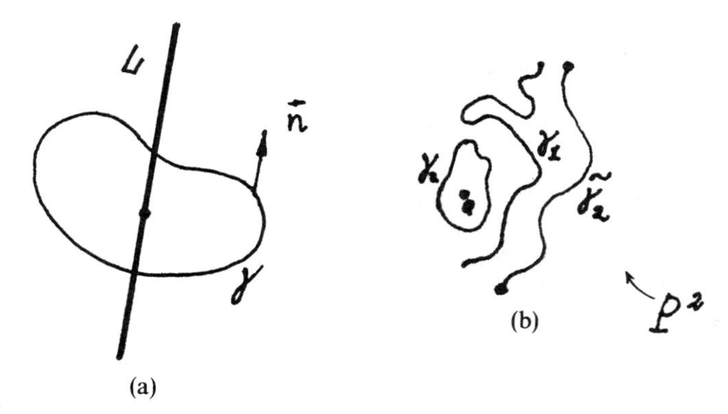

Figure 12.2: Disclination in a nematic crystal. (a) L is a thread; γ is a contour surrounding a thread. (b) γ_1 is the image of the thread γ under a mapping into the order parameter space P^2; $\tilde{\gamma}_2$ is homotopic to a noncontractible curve on the projective plane. The curve γ_2 contracts to point a.

Both "permissible" cases occur in nature. In the classical experiments performed in the early 1920s by Grandjean and Friedel a system of critical points was observed on the surface of a nematic. When viewed through crossed Nicol prisms the critical points appear as joined black bands. Friedel called these points *nuclei*, and he proposed a remarkably simple way of distinguishing types of nuclei. If a nucleus is observed on the surface between the nematic and the covering glass, the glass is moved in its plane. When a nucleus corresponds to the end of a vertical line inside the nematic, the line is bent and looks like a black thread. A nucleus forms as the ending of a singular thread in the case of a stable disclination; an unstable disclination disintegrates in the model. This is completely natural from a topological point of view.

The situation is even more complex in cholesteric liquid crystals. Locally—that is, at distances of the order of the length of molecules—a cholesteric is similar to a nematic. There is also no regularity in the location of the centers of the molecules, but the molecules themselves are oriented along an axis directed according to the director **n** (the local optical axis). In contrast to the nematic, however, a different type of oriented orderliness arises in the cholesteric phase. The state of the cholesteric is not fixed by the direction of the director **n**. The director **n** varies continuously in space, describing a heliacal curve. If one somewhat schematically represents a cholesteric as a collection of flat molecular layers, then given the transition from layer to layer, the vector **n** rotates continuously, describing a spiral. This approach does not take into consideration the possibility of the molecules rotating to exit from the "layer." But for our purposes, such a simplification is completely justified. If one looks at the monocrystal of a cholesteric in a layer with thickness on the order of 100 microns, then given exposure to light, we distinctly catch sight of a spiral structure. Release a ray of light along the axis of a spiral. The reflected light is circularly polarized. At each instant the picture of the electric field in the reflected wave is similar to a spiral, identical in form to the spiral of a cholesteric. If we look at a cholesteric in a layer caught between two flat discs and assign tangential boundary conditions, then the director **n** will be defined by simple formulas:

$$n_x = \cos(q_0 z + \rho), \quad n_y = \sin(q_0 z + \phi), \quad n_z = 0.$$

The axis of the spiral is directed along the z-axis, q_0 is the wave vector, and the angle ϕ is determined by the boundary conditions (see Fig. 12.3). The spatial period L of the spiral is equal to half a step of a spiral (because the vectors **n** and $-$**n** are equivalent):

$$L = \frac{\pi}{|q_0|}.$$

For the majority of cholesterics, the magnitude of L is of the order of 3000 Å. This distance is significantly greater than the length of the molecules and is comparable to the length of a light wave, which explains the possibility of visual observation of the spiral structure in a cholesteric.

Let us now turn to a study of the structure of critical points in cholesterics. Just as in nematics, the problem is reduced to a study of the topology of the space of the order parameter—the parameter of the degeneration of the system.

At the outset let us define an order parameter. In essence, this was already done for a cholesteric included in the space between two flat disks. In general the order parameter is characterized by the reference frame, consisting of a set of three vectors: $\mathbf{n}, \mathbf{d}, \mathbf{l} = \mathbf{n} \times \mathbf{d}$. Here **n** is the director, **d** is the unit vector directed along the axis of the spiral, and **l** is their cross product. To find the full domain of variation of the order parameter one must also consider the additional symmetry of the system relative to the transformations $\mathbf{n} \mapsto -\mathbf{n}, \mathbf{d} \mapsto -\mathbf{d}, \mathbf{l} \mapsto \mathbf{l}$.

How does one find threadlike and point singularities in cholesterics? The homotopy techniques we discussed in the case of a nematic turn out to be very effective in studying critical points in cholesterics, but, of course, they are complicated by the nontrivial structure of the order parameter.

Consider first the problem of classifying point and line singularities in a volume of cholesteric liquid. The solution of this problem is completely known. There are no point defects in a volume of cholesteric liquid. The classification of line singularities turns out to be significantly more subtle. Threadlike singularities in a cholesteric cannot be determined by one integer parameter. They are defined by elements of a finite group—the group of unit quaternions. The quaternion group is a natural extension of the group of complex numbers. Each complex number can be represented as $z = x + iy$, where $i = \sqrt{-1}$. The numbers 1 and i are called *generators* in the group of complex numbers. The nonzero complex numbers form a group whose group operation is multiplication. If the generating elements 1 and i are supplemented by elements -1 and $-i$, and the group operation is taken to be multiplication, the result is a new group—a finite group of four elements. This group is commutative, that is, the

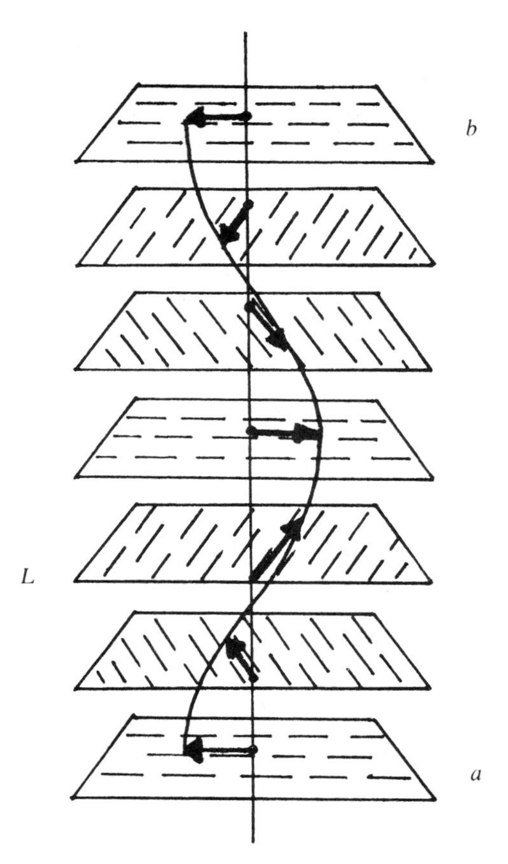

Figure 12.3: A cholesteric contained between two disks *a* and *b*. A schematic representation of a helical structure. (Interstitial planes are drawn for a more visual depiction of the change of the director and do not have real physical meaning.)

product of any two elements does not depend on the order of the factors. For the group of quaternions, the generators will be elements 1, **i**, **j**, **k** which are completely defined by assigning the rules of multiplication among themselves: $\mathbf{i}^2 = \mathbf{k}^2 = \mathbf{j}^2 = -1$, $\mathbf{ij} = -\mathbf{ji} = \mathbf{k}$, $\mathbf{ki} = -\mathbf{ik} = \mathbf{j}$, $\mathbf{jk} = -\mathbf{kj} = \mathbf{i}$. The group of quaternions, as shown by the multiplication rule, is noncommutative: the result of multiplying two elements depends on the order of the terms.

It turns out that in the case of a cholesteric the group responsible for the appearance of threadlike singularities will be precisely this group of unit quaternions—the fundamental group of the space of the order parameter of a cholesteric. Its noncommutativity shows up in the classification of critical points. For example, in the process of adhesion there is an order in the splicing of two distinct singular threads.

A new class of topological problems arises in the investigation of boundary effects in a cholesteric. In this case it is useful to apply the techniques of homology theory.

Imagine that a cholesteric liquid fills a region bounded by a closed two-dimensional orientable surface M^2. One can then ask if vortices or critical points exist on the surface of a cholesteric. Let us assign boundary conditions on the surface such that the vector **d** is normal to the surface, while the vectors **n** and **l**, which are orthogonal to the vector **d**, lie in the tangent plane. The vectors **n** and **l** form a tangent field to the surface. If on the surface one can define a field of tangent vectors that depends continuously on the points on the surface and nowhere vanishes, it is obvious that vortices of the vector field **d**—critical points on the surface M^2—cannot arise.

The answer to the question posed is found in Poincaré's hedgehog theorem. In fact, that theorem immediately implies that the only manifold on which there exists a vector field without critical points is a torus. In all other cases, there must be critical points on the surface.

So far we have solved only part of the problem. One would like to know in more detail the structure of the critical point on a surface. Poincaré's theorem makes it possible to obtain several bounds on the number and form of vortices on the surface. As we already know, the formation of a vortex in the vector field **d** corresponds to the occurrence of a critical point in a tangent field. An integer—the *index*—characterizes each critical point a. To define the index of a critical point on a surface, one proceeds as follows. Let a be a critical point. We enclose the point a with a sufficiently small circle γ so that the neighborhood which is obtained is planar. We orient the circle γ in a counterclockwise direction and examine the change of the vector field $\mathbf{n}(x)$ after one circuit of the circle. The vector **n** defined on the circle will make, in this case, a certain whole number of rotations k. This number is called the *index* of the vector field at the critical point a.

The number k can be negative or positive depending on which direction vector **n** moves as the point moves along the circle. If the direction of movement of **n** coincides with the orientation of γ, then k is positive; if not, then k is negative. Physicists attach a different form to the concept of an index and talk about quantized vortices of the vector field normal to the surface. In physical terms the index k of a vector field is the force of a quantum vortex or the number of quanta of circulation (Fig. 12.4).

In its full generality, the formulation of Poincaré's theorem is as follows: *If a field of nonzero tangent vectors defined on a surface M^2 is continuous everywhere except for a finite number of critical points, then the sum of indices of all the critical points is equal to the Euler characteristic of the surface $\chi(M^2)$.*

This theory immediately yields a number of restrictions on the types of surface singularities. For example, the Euler characteristic of a doughnut with two holes C^2 is $\chi(C^2) = -2$. One obviously cannot construct vortices having only positive circulation on this surface. The opposite situation occurs on a sphere; here $\chi(S^2) = 2$.

From a topological point of view, the existence either of two vortices with circulation strength 1 or of one vortex with $k = 2$ is possible. Topology cannot determine which solution really exists. One can answer this question by considering the energy of the system of vortices. It turns out that a system of one vortex with two quanta of singularity requires less energy (Fig. 12.5). The bending of the bound state of a system with two vortices is higher than that of a vortex with two quanta of circulation.

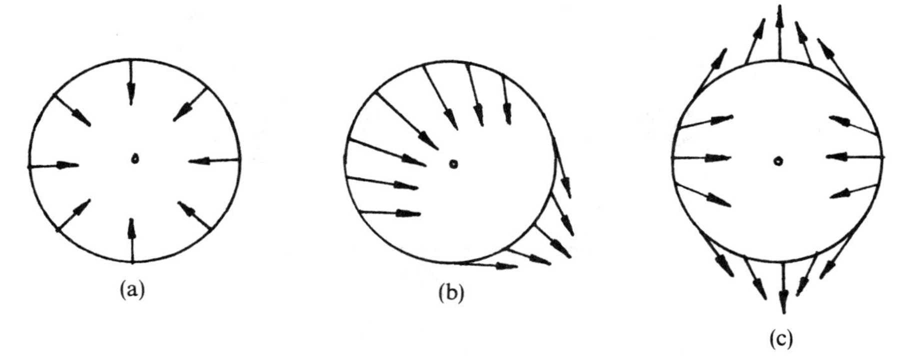

Figure 12.4: The index of a vector field. (a) $k = 1$. (b) $k = 0$. (c) $k = -1$. If the index k equals 0, then the vector field does not have critical points.

Since topology prohibits the existence of an isolated vortex with circulation strength 1, two isolated critical points on a spherical surface can appear only as the ending of singular lines inside an object (see Fig. 12.5b). It is clear that the surface singularities cannot be the ends of the only vortex line since in this case they have opposite signs of circulation.

It has already been noted that a torus is the only configuration on which a field without critical points is possible. But this does not completely exclude the formation, for example, of two vortices with opposite index values. These vortices should cancel each other.

The appearance of boundary critical points is characteristic of many ordered media. Such solutions were first discovered in one of the phases of superfluid ^{3}He, the A-phase, whose order parameter space is topologically close to that of the cholesteric phase. The existence of such surface-point vortices leads to striking effects of fundamental significance in the theory of ordered states. One of them is the decay of a superfluid current on a surface when there is a collision with a similar singularity.

Given physicists' characteristic love of neologisms, this phenomenon has been called by a number of names. The most widely accepted term, *boojum*, very aptly characterizes the enigmatic properties of critical points. The American physicist D. Mermin proposed this term, having borrowed it from Lewis Carroll's poem "The Hunting of the Snark."[1] (Those hunting for the mysterious Snark are cautioned to be careful lest the Snark turn out to be a boojum. Just what the "suddenly vanishing boojum" was, remains an enigma.)

In investigating the topological structure of critical points in cholesteric crystals, we implicitly assumed the existence of boundary conditions, fixing the vector **d**

[1] In the article, "E Pluribus Boojum," published in the April 1981 issue of *Physics Today* (p. 46), Mermin has many interesting things to say about the history of the term *boojum* and about the humorous adventures met in trying to gain "citizenship" for the term among physicists.

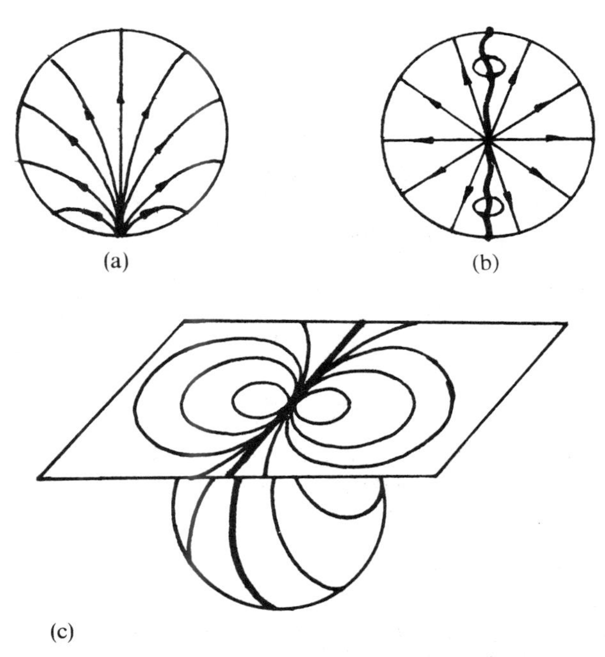

Figure 12.5: (a) Vortex with $k = 2$—a boojum. (b) Two vortices with $k = 1$. (c) A vector field on a sphere having one critical point at the north pole (N) and consequently an index of 2.

normal to the surface. The possibility of choosing such boundary conditions requires additional considerations because, at first glance, the constant pitch of the helix prevents this. Such a problem does not arise for the A-phase superfluid ^{3}He, where the reference frame of the order parameter has another physical meaning. Since the choice of a cholesteric as an example of the applications of topology was due to its greater accessibility, I shall not dwell on these questions.

In a narrow range of temperatures (0.1–1 K) one of the most interesting and mysterious liquid crystals lies between the cholesteric and isotropic liquid—the so-called *blue phase*, more precisely three varieties of blue phases—phases I, II, and III. The blue phase was discovered by the original discoverer of liquid crystals Fridrich Reinitzer (1857–1922) almost simultaneously with his observation of nematics and cholesterics, but it remained a little-studied object for a long time. The blue phase takes its name from its remarkable optical properties. When it is illuminated with polarized light, a selective scattering (Bragg reflection) is observed in the visible part of the spectrum, primarily in the blue range (which does not preclude seeing orange or red light). The presence of Bragg reflection indicates a certain periodicity in the structure of the blue phase. And indeed, the blue phase combines spatial periodicity with local cholesteric behavior. However, in contrast to the cholesteric, in which the director describes a spiral, the tensor order parameter of the blue phase, which characterizes the correlation in the orientation of molecules, has spatial periodicity. This ambiguity leads to a significant complication of the structure of the blue phase and in particular in the structure of the defects that form the doubly periodic system. While a description of types I and II of the blue phase can be obtained in the context of the modified Landau theory, in no way can this be said of phase III, which is called the "fog" phase, and cannot be incorporated in the general theory. In regard to its properties the "fog" phase is closer to an isotropic liquid, but may be connected with a weaker symmetry of icosahedral type. If this is confirmed, we shall have obtained an exotic example of a liquid quasicrystal. Realizations of the most arcane geometric objects seem to be possible in the blue phases, in the form of textures and defects.

The other class of liquid crystals, smectics, is characterized by a constant distance between layers that naturally prevents the application of homotopy methods because the contracting of contours is possible only in separate layers.

The French mathematician V. Poenaru has suggested using the theory of foliations for classifying singularities in foliated media. The theory of foliations, which studies the qualitative structure of hypersurfaces with attached vector fields, is a comparatively new branch of mathematics. It includes as a special case the qualitative theory of differential equations; but so far there have been too few concrete applications of the theory of foliations to mention them in popular literature.

Yet another area of application of topology has been discovered comparatively recently. The experimental discovery of the superfluid state of ^{3}He was an important event of the 1970s. The possibility that ^{3}He could make the transition to a superfluid state was predicted in 1959 by the Soviet physicist L.P. Pitaevskii, but it was only twelve years later that the American physicists D. Osheroff, R.C. Richardson, and D.M. Lee succeeded in obtaining superfluid liquid ^{3}He at fantastically low temperatures, on the order of 2×10^{-3} K. The scientists who made this remarkable discovery were awarded the Nobel Prize twenty-five years later (in 1996). The appearance of superfluidity in ^{3}He is connected with the same effects of Cooper pairing of the Fermi system that one finds in superconductivity; but whereas in superconductivity Cooper pairs have a full spin of 0, in ^{3}He the spin equals 1. A striking property of superfluid ^{3}He, detected experimentally and also predicted theoretically, was the appearance of several phase transitions. At temperatures of the order of 2.6×10^{-3} K, ^{3}He will be in one thermodynamic phase, the A-phase, but given a further lowering of temperature to 2.1×10^{-3} K, it changes to a different phase—the B-phase. These phases possess different properties. For example, the A-phase is anisotropic while the B-phase is isotropic.

It is quite significant that in a neighborhood of the points of a phase transition the equations of a state admit a macroscopic description by means of equations analogous to the Ginzburg–Landau equations in the theory of superconductivity.

It was shown that a series of important characteristics of the superfluid phase of ^{3}He follow from topological and group considerations of the space of the order parameter. The order parameter is given by a complex 3×3 matrix. The dimension of the order parameter space is 18. This space contains the domains of variation of the order parameters of the A-phase and B-phases, which have dimensions 5 and 4, respectively. In the context of a phenomenological description of superfluid ^{3}He, a change of the topological properties of the thermodynamic A- and B-phases is observed. In this situation additional fields are used, in particular, a magnetic field.

Point defects arise in the superfluid phases of ^{3}He. They are classified by using the same methods that apply in the theory of liquid crystals, but, of course, they are complicated by the higher dimensionality of the order parameter space. Problems of another type arise with ^{3}He. For example, it is not yet clear how many distinct thermodynamic phases there can be. Thus far three phases have been found. Besides the A- and B-phases, an A_1-phase was recently detected in a powerful magnetic field. Many physicists have attempted to make a theoretical estimate of the number of phases in ^{3}He. F.A. Bogomolov and the present author, using a special group-theoretic construction, have proved that the maximum possible number of phases is 11. Our method develops an approach of V.L. Golo and the author, who showed that the phases can be associated with the orbits of the free energy potential—the Ginzburg–Landau potential. The advantage of the proposed construction is that it gives both a proof of the completeness of the resulting classification and an explicit description of the discrete invariants of the phases (the discrete subgroups of the stationary groups of orbits), which characterize the topology of the defects in ^{3}He. Of course, only experiment will reveal which of the phases thereby found are realized in nature. These questions are discussed in detail with citations of the original works in the book [Mo].

Very interesting problems arise in the fluid dynamics of superfluid ^{3}He. In essence, we are lacking a complete system of equations to describe the behavior of superfluid liquids in the presence of vortices and other singularities, the existence of which follows from topology.

One can obtain more detailed results for steady-state solutions—those which do not depend on time. Recently, the problem of spatially one-dimensional steady-state solutions in the A- and B-phases was solved. It was shown that the corresponding equations are equivalent to equations of a mechanical top with variable axes of inertia. Such solutions describe phenomena which occur in very narrow capillaries filled with superfluid ^{3}He.

The last ten or fifteen years have seen an intensive study of vortices in superfluid ^{3}He. In contrast to ^{4}He, vortices in ^{3}He have a very large core, in which different phases can coexist. Oversimplifying a bit, one might say that in the core of a vortex the same processes occur as in the bulk ^{3}He. The theoretical analysis of the structure of vortices is based on a combination of topological ideas and analogues with quantum field theory. Essentially ^{3}He is a complicated nonlinear chiral field model with a multidimensional order parameter concealing deep mathematical and physical structures. A remarkable characteristic of studies of ^{3}He is the unity of theory and experiment, which is rare in our time. On the cryostat built in low-temperature laboratory at the University of Helsinki experiments have been conducted on rotating ^{3}He. Very exotic vorticial states have been observed: for example, vortices in the B-phase with half-integer quanta of singularity and vorticial surfaces. These phenomena had been predicted theoretically, and the experimental results in turn were a powerful impetus to the development of the theory. Experimental and theoretical studies of ^{3}He being conducted in the USA, Finland, France, Russia, and a number of other countries will undoubtedly yield new discoveries of fundamental value.

Very promising applications of these methods have been discovered recently in astrophysics during investigation of the structure of pulsars, white dwarf stars, and other superdense stars. A well-known example is a neutron star. The nucleus of a neutron star, given pressure of the order of 10^{14} g/cm^3 and temperature $T = 10^9$ K, is in a superfluid state. The properties of superfluid nuclei resemble those of superfluid ^{3}He.

It is natural to expect that methods developed in the theory of ^{3}He will prove useful also in this new area. It is especially interesting to study the structure of vortex solutions in the nucleus of a neutron star. The formation of vortices in a rotating neutron star leads to interesting observational effects. One possible explanation of the fluctuation of the period of the pulsar in the Crab Nebula is related to this phenomenon. Astrophysicists identify this pulsar as a neutron star. A lattice of vortex threads forms in a rotating neutron star, while (as the Soviet physicist V. Tkachenko has shown) an acoustic wave should form in such a system and propagate in a plane perpendicular to the vortex threads. The period of the Tkachenko wave, calculated for the pulsar of the Crab Nebula, agrees well with the characteristic parameters of a neutron star.

Topological methods entered the realm of physics known as the theory of ordered media in June 1976, in the letters section of the French *Journal de Physique*. The journal contained a short note by the French physicists M. Kléman and G. Toulouse, entitled "Principles of a classification of defects in ordered media." This note contained their announcement of results on the classification of line and point defects in ordered media—in particular, they considered nematics and the A-phase of superfluid ^{3}He using homotopy methods. This work initiated feverish activity in applying topology to condensed-matter physics.

The universal and instant recognition of the effectiveness of topological methods has elicited a certain surprise. The marked conservatism among physicists in regard to the application of new, or more precisely, unaccustomed mathematical methods is well known. In addition, a large part of the results obtained were already known to physicists in one form or another. For example, the classification of singularities in cholesterics, had been obtained earlier by Kléman and J. Friedel (grandson of G. Friedel). In essence, the classification of threadlike singularities, achieved in the 1920s using the Frank index, is purely topological.

Nonetheless, the precise understanding of the topological nature of the occurrence of defects and the application of corresponding mathematical techniques have led not only to a refinement of the former classification schemes and the removal of several gaps, but also to new results in systems of ^{3}He type with multidimensional degeneration of the order parameter.

In this chapter I have been able to give only the most general idea of applications of topology. In essence we are talking about using two principles of topology: homotopy theory and homology theory. The time has now come for more subtle applications. For example, the methods of Thom's catastrophe theory may prove to be useful in studying the interaction of several singularities in cholesterics or in the description of the caustics that arise in the irradiation of patterns and defects in nematics [JR], [JMR]. In the physics of ordered media there remain many unsolved problems which are interesting from the point of view of topology. Our next chapter, however, will deal with a different area of physics, quantum field theory.

The significant application of homotopy theory began two years earlier. Perhaps this provides a psychological clue to the recognition of topology in the physics of solids.[2]

---

[2]At the end of the article cited above, Toulouse and Kléman wrote that they had discovered a use of analogous methods in field theory in the 1974 work of Perelomov and the author on classifying monopoles of 't Hooft–Polyakov (see the chapter "Topological Particles") and in the 1959 work of D. Finkelstein and C.W. Misner about conservation laws in gravitation theory.

Chapter 13

Theory of Gauge Fields

THE basis for all contemporary theories of elementary particles is the concept of a field. For each elementary particle there is a corresponding quantum field, and the interaction of elementary particles is defined by the interaction of quantum fields.

The quantum theory of light is an example of the most consistent embodiment of these ideas. The concept of light as a beam of elementary particles, of photons, can be combined with the wave picture, by imagining that a free electromagnetic field is quantized. The energy of an electromagnetic field is equal to the sum of the energy of the elementary "pieces" of the field—electromagnetic field quanta. Electromagnetic field quanta are photons. A quantitative relation between the energy of a photon ε and the frequency ν of the corresponding field is given by the well-known Planck–Einstein formula $\varepsilon = \hbar\nu$ where $\hbar$ is the Planck constant.

The quantization procedure gives meaning to the concept of a single photon. One can talk about processes of emission and absorption of photons. For example, one can present the interaction of two electrons as a photon exchange process. Thus, the first electron "emits" a photon, while the second electron "absorbs" it.

A broad range of phenomena exist where the interaction of elementary particles with light plays a decisive role. Such interactions are called electromagnetic. The theory of these processes constitutes the subject of quantum electrodynamics. A quantum description is possible due to the relatively small magnitude of electromagnetic interactions. This magnitude is described by the fine structure constant, usually denoted by α. The constant α is dimensionless and equal to $e^2/\hbar c$; its value is 1/137.

The basic mathematical apparatus of quantum electrodynamics has become perturbation theory, which makes it possible to calculate all physical quantities by expansion in a series of powers of α.

The idea of quantum electrodynamics has been triumphantly confirmed by many experiments. It suffices to mention the measurement of the magnetic moment of an electron. Its theoretical value coincided with the experimental value with an accuracy of one hundred-thousandth of one per cent. The successful application of the perturbation theory in quantum electrodynamics did not increase the interest of active physicists in the more rarified areas of mathematics. A situation developed analogous to that in quantum mechanics in the 1930s. The Swiss physicist R. Jost wittily characterized it:

> In the '30s, under the demoralizing influence of the quantum-mechanical theory of perturbations, the need of theoretical physicists for mathematical knowledge came down to a rudimentary mastery of the Latin and Greek alphabets.

As with any good joke, this witticism should not be taken literally. The techniques of Feynman diagrams, underlying the invariant theory of perturbations, contains mathematical profundity, but its strength is that it makes it possible to obtain fundamental results in physics without being entangled in a morass of mathematical reasoning.

Two "clouds" still remained in elementary particle physics, however. Besides electromagnetic interactions, two other types are distinguished: weak and strong interactions. In contrast to the long-range electromagnetic field, weak and strong interactions are short-range. The range of strong interaction is approximately 10^{-13} cm. This is the order of the diameter of a strongly interacting particle. If one takes the magnitude of strong interaction at this distance as a unit, then the electromagnetic interaction is 137 times weaker than a strong one while a weak interaction amounts to only one millionth (10^{-6}) of a strong interaction. There also exists a fourth force (gravity), the weakest of all, with magnitude 10^{-39} times the magnitude of a strong interaction. It acts on the scale of the universe as a force of attraction. The whole complex world of elementary particles, including the processes of scattering, creation, annihilation, and transformation of particles into one another is governed by the four fundamental forces.

The distinction of interaction types, in particular the classification into strong and weak interactions, is largely a matter of the magnitude of the interaction and the nature of the physical processes. A force of interaction is determined by characteristic scales and invariant charges (strong) g_s and (weak) g_w. These charges play the role of coupling constants, but depend in general on the distances between the particles, or, what is the same, on the momenta q transmitted.

As q increases (that is, the distances decrease), the quantities g_w and g_s decrease, and at certain values $q \sim 10^{15}$ GeV they become equal. As q increases further the charges g_s and g_w tend asymptotically to zero. At the same time the constant of electromagnetic interaction (electric charge) e increases and at the same energy (10^{15} GeV) all three charge invariants are equal. Thus at this energy one can speak of the unification of all forms of interaction. But at this energy level the gravitational interaction is still too weak. It becomes the determining factor at energies $\sim 10^{19}$ GeV. This scale is referred to as the *Planck scale*.

Unfortunately analysis of gravitational interaction in this case precludes any experimental test under laboratory conditions. No future accelerators, much less any present accelerators (which attain maximum energies of 10^3 GeV) will even approximate the Planck energy. For that reason the only "experimental foundation" for testing theoretical conceptions of the behavior of particles on the Planck scale lies in astrophysics. The theory of the origin of the early Universe (the Big Bang) predicts that in the first few instants when the Universe was being formed its temperature was on the order of the Planck scale.

A physicist's fondest wish is to construct a unified theory of elementary particles that will include all types of interactions. At present, no one will predict when such a theory will be created, but impressive successes have been achieved in this direction.

It would have been very difficult to progress if it were necessary to consider the effects of all forces at once, but Nature has met us halfway. It is quite clear that on the nuclear scale the influence of gravitational forces is completely negligible, and therefore one can disregard them at first. A second happy circumstance is the relative independence of weak and electromagnetic interactions, on the one hand, and strong interactions, on the other. A whole class of elementary particles exists which participate in weak and electromagnetic interactions but not in strong interactions. Such particles are called leptons, examples of which are the neutrino, electron, and μ-meson. The majority of well-known elementary particles participate in strong interactions, for example, the proton, the neutron, the Σ-particle, the recently discovered J/ψ-particles and the K-, π-, and ρ-mesons. Strong interactions are sometimes called nuclear because they determine the bound states of neutrons and protons in a nucleus. It does not follow that elementary particles taking part in various processes remain entirely distinct. For example, a proton participates in strong interactions, but to the extent that it is electrically charged, it reacts to an electromagnetic field. It also appears in the β-decay of a neutron—a weak interaction. There is a strict distinction in physics between the theory of strong interactions on the one hand, and weak and electromagnetic on the other, which, in a first approximation, makes it possible to study these processes separately.

Nonetheless, all modern theories of elementary particles share one common idea—the concept of gauge invariance.

The classical example of a gauge theory, which has existed for more than one hundred years, is Maxwell's theory of electromagnetism. One can write the Maxwell equations using a four-dimensional vector potential A_μ. It turns out that the equations do not change if we add to A_μ the derivative of an arbitrary function of the coordinates of four-dimensional space-time:

$$A_\mu \mapsto A_\mu - \partial_\mu \eta(x) \quad (\mu = 1, 2, 3, 4).$$

This property is called the gauge invariance of Maxwell's equations. It plays a fundamental role in electrodynamics, placing strict limitations on the form of the equations of motion.

Gauge transformations of the field A_μ form a group—the *gauge group* of equations of electrodynamics.

It is known that conservation laws are connected with the presence of a symmetry group. Global symmetry groups, which act on the whole space, are distinguished from groups generated by transformations in a neighborhood of some point, which are called local symmetry groups. The former are defined by global conservation laws: conservation of energy, momentum, and so forth. The latter are defined by the invariants of gauge fields. For example, conservation of electric charge for an interaction with a gauge field A_μ is associated with the gauge transformations mentioned above. In electrodynamics the charge of an electron e plays a dual role. The conservation of electrical charge in electromagnetic interactions limits the types of reactions of elementary particles; at the same time, it plays the role of a single "coupling constant," that is, it defines the magnitude of the interaction. Moreover, assuming that the interaction between particles is defined by a gauge field and assuming an electric charge, all the remaining characteristics of particles, for example, magnetic moment, are consequences of the equations of motion.

In 1954 the American physicists C.N. Yang and R.L. Mills proposed a model of strong interactions, introducing a new class of gauge fields. The Yang–Mills theory was originally intended to describe forces binding nucleons—protons and neutrons. Protons and neutrons possess a remarkable peculiarity that makes it possible to regard them as identical in strong interactions. This property had been observed earlier in the context of quantum mechanics and was called *isotopic invariance.* A proton p and a neutron n are two states in which a nucleon can be found. One state, the proton, has a positive electrical charge; the other, the neutron, is neutral.

For the characteristics of the two distinct "charged states" p and n of a nucleon, a special vector τ is introduced. The projection of the vector τ onto a fixed axis in three-dimensional space admits only two values: $+1/2$ and $-1/2$. By convention the value $+1/2$ is assigned to a proton and $-1/2$ to a neutron. The vector τ is called the *isospin vector* because by its properties it is similar to the usual spin $1/2$, but it has a completely different physical nature. The isospin vector lies in an auxiliary space created by the wave functions of the proton and the neutron—the space of states of a nucleon. This is the isotopic spin space.

The space $\mathbb{C}^2$ contains the transformation group

$$\begin{aligned} |p'\rangle &= \alpha_{11}|p\rangle + \alpha_{12}|n\rangle \\ |n'\rangle &= \alpha_{21}|p\rangle + \alpha_{22}|n\rangle. \end{aligned}$$

Here the α_{ij} are complex numbers satisfying additional relationships:

$$\alpha_{ik}\alpha^{\dagger}_{kj} = \delta_{ij}, \qquad \det(\alpha_{ij}) = 1.$$

It is called the isotopic group of the nucleon or simply the isospin group. The standard notation for it is SU(2). Transformations from the group SU(2) do not change the state of a nucleon. This is the definition of isotopic invariance.

The conservation of isotopic invariance is characteristic of strong interactions, during which proton-neutron (pn), neutron-neutron (nn) and proton-proton (pp) systems appear completely identical. Isotopic symmetry is broken only in cases of electromagnetic interactions. We recall that a proton is positively charged, while a neutron is electrically neutral. This leads to a small difference in mass. The mass of a neutron is $\sim 939.6\,\text{MeV}$, and is $1.3\,\text{MeV}$ larger than the mass of a proton ($1\text{MeV} = 10^{-6}$erg.) Isotopic invariance is one of the internal symmetries of elementary particles because it is not connected with the kinematics and dynamics of nuclear reaction.

A whole series of other quantum numbers are known—quantities that are conserved in strong interactions and which have the character of internal symmetries. For example, an important characteristic, which separates strongly interacting elementary particles into two classes, is the baryon number or baryon charge A, analogous to the mass number of a nucleus known in atomic physics. The mass number is the number of neutrons and protons comprising a nucleus. For neutrons and protons, the baryon charge A is 1. A particle with $A = -1$ is called an *antibaryon*. Mesons have $A = 0$. The conservation law of baryons asserts that the total value of A does not change in any process. Baryons cannot be destroyed or created, with the exception of the processes of the annihilation of a baryon and an antibaryon or the creation of a pair.

Yang and Mills had a bold idea. They suggested that certain internal symmetries have a local, as well as a global, character.

In their 1954 paper Yang and Mills showed that the properties of nuclear forces connected with isotopic symmetry and consequently, together with the indistinguishability of a neutron and a proton, are preserved under much more general transformations. Transformations of the isospin vector can occur independently at each point of space-time.

These transformations form a group isomorphic to the group of transformations of isospace, but they act in a neighborhood of each point of space-time. This group, the local isospin group, is noncommutative. The property of noncommutativity turns out to be very significant in the development of the theory.

With this background, one can now explain the Yang–Mills scheme: strong interactions are constructed in analogy with electrodynamics, but taking account of the specific characteristics of a noncommutative gauge group, the isospin group.

An analog of the gauge field A_μ for strong interactions will be the gauge field W_μ defined by properties of gauge invariance of equations relative to the local isospin group. Just as in electrodynamics, the interaction can be pictured as an exchange of massless particles. But in the case of electrodynamics, the gauge field is the electromagnetic field A_μ and the particles participating in electromagnetic interaction exchange a photon. In strong interactions the massless field W_μ is a vector in relation to isospin transformations. Yet another important difference between the Yang–Mills field W_μ and an electromagnetic field is connected with the noncommutativity of the isospin group. An electromagnetic field interacts directly only with charged external particles, but in a Yang–Mills field different components of the field interact with one another.

All these physical considerations are realized in the Yang–Mills Lagrangian, more precisely, the density for the Lagrangian $\mathcal{L}_{\mathrm{YM}}$, which we write out for the simplest isospin group SU(2). We denote the Yang–Mills field intensity by $F^a_{\mu\nu}$ (where $\mu, \nu = 1, 2, 3, 4$ are spatial indices and $a = 1, 2, 3$ are isotopic indices). Then

$$F^a_{\mu\nu} = \partial_\mu W^a_\nu - \partial_\nu W^a_\mu + g[W_\mu, W_\nu].$$

Here g is the coupling constant, $[W_\mu, W_\nu]$ is the commutator of the fields,

$$[W_\mu, W_\nu] = \varepsilon_{abc} W^b_\mu W^c_\nu.$$

(Here and below, summation over repeated indices is assumed.) $W_\mu = W_\mu^a t^a$, where t^a are the generators of the Lie algebra of the group SU(2). They can be expressed in terms of the Pauli isospin matrices. We give the density $\mathcal{L}_{\mathrm{YM}}$ in the form

$$\mathcal{L}_{\mathrm{YM}} = -\mathrm{Tr}\,(F_{\mu\nu}F_{\mu\nu}). \tag{13.1}$$

This scalar quantity is invariant relative to the gauge transformations $F_{\mu\nu} \mapsto gF_{\mu\nu}g^{-1}$, $g \in \mathrm{SU}(2)$, and is called (the density of) the *Yang–Mills Lagrangian*. The Lagrangian $\mathcal{L}_{\mathrm{YM}}$ can also be easily written out for an arbitrary gauge Lie group G by replacing the generators t^a with the generators of the Lie algebra of the group G.

The work of Yang and Mills did not receive immediate recognition. The principle of gauge invariance is easy to realize in electrodynamics. The massless particle associated with the gauge field is the photon. In practice it is easy to verify its existence by glancing through a window or turning on a light. However, other vector massless particles do not seem to occur naturally. Therefore, the majority of pragmatically minded physicists did not take the Yang–Mills theory seriously. Ten years passed before events in the world of elementary particles forced all theoreticians to recall the work of Yang and Mills.

In 1964 two theoretical discoveries were made. One of them was called quark theory; the other was called the "Higgs effect" or the "Higgs mechanism." Quark theory acquired a sensational reputation. The world's largest news agencies made announcements about quarks, major articles in journals like *Newsweek* were devoted to them, ballet pantomimes were staged, and so forth. The Higgs effect became the property of only a narrow circle of specialists, but after only a few years theories appeared in which the Higgs mechanism rendered a major service to "quark" theory.

Quarks were invented independently by two physicists, the American theoretician M. Gell-Mann and the Swiss G. Zweig.[1] Attempting to introduce order into the more-than-slightly confused world of strongly interacting particles, Gell-Mann and Zweig proposed that all particles participating in strong interactions—called hadrons—are compound particles. In the original Gell-Mann/Zweig scheme, there were three fundamental particles, q-{u(up, d(down), s(strange)} quarks. It was suggested that quarks (q) are the components of all observed particles. More precisely, in order to represent all particles it is necessary to introduce antiquarks $\bar{q}$ as well. Then all baryons B consist of three quarks: $B = qqq$, $\overline{B} = \bar{q}\bar{q}\bar{q}$, while mesons consist of quark/antiquark pairs: $M = q\bar{q}$.

Quarks must have very unusual properties; for example, they have a fractional electrical charge. Attempts over many years to observe free quarks have not been successful. It is true that occasionally sensational reports have appeared in print about the detection of particles with fractional electrical charge, but more thorough verification has not supported these observations. A natural question arises: *Do free quarks exist?* At present the answer is approximately as follows: They are too important for physicists to get along without them.

---

[1]The name *quarks* for these hypothetical particles was proposed by Gell-Mann, who borrowed it from James Joyce's *Finnegan's Wake* (New York: Viking Press, 1939), which contains the puzzling line, "Three quarks for Muster Mark!" (p. 383). The term *aces*, invented by Zweig, did not catch on.

All contemporary models of strong interactions are based on quark models, but the original picture of their properties has undergone considerable change. The discovery of a large number of new hadrons has led to a significant complication of the theory. Three quarks are not sufficient to construct all the new particles that have arisen. One must introduce yet another—a fourth quark. It is called *charm*. The latest schemes already involve a fifth quark, and it is likely that this is not the end of the matter.

The J/ψ mesons consist precisely of $c\bar{c}$-pairs. The discovery of heavier baryons and mesons with explicit charm of type $c\bar{u}$ required the introduction of new quarks called b-(bottom-), and t-(top-) quarks. A spectacular confirmation of the quark model was the successive experimental discovery of additional quarks. The last t-quark was found in 1995, in experiments performed by two groups of researchers: the CDF and D0 collaborations on the Tevatron at the Fermilab in Batavia, Illinois. It turned out to be very heavy, with mass $\sim$ 1756 GeV. It is no wonder that the detection of the top-quark is possible only on the most powerful accelerator in the world (the $p\bar{p}$-collider, with a center-of-mass energy of about 1800 GeV. Thus, at present all 6 quarks predicted by theoreticians have been detected experimentally: $\begin{pmatrix} u & c & b \\ d & s & t \end{pmatrix}$.

The generally recognized current model of strong interactions is quantum chromodynamics. The very name contains an analogy with quantum electrodynamics. The fundamental ideas of this theory will become clearer after a more detailed classification of quarks. The terminology in the field of quarks is exceptionally colorful. The "modern" quark q^i_j is a complex particle. It possesses two degrees of freedom. One degree of freedom is associated with isotopic parameters. The number of indices i is determined by the number of quarks. The corresponding states are called *flavors*. A second degree of freedom is called *color*. Each of the Gell-Mann/Zweig quarks, p, n, λ, can be found in three color states forming a colorless hadron. Originally Gell-Mann proposed coloring quarks with the colors of the American flag: red, white and blue. The set of colors that eventually became generally accepted corresponds to the three primary colors of visible light: red, blue and yellow. Their combination produces white light. Therefore, three colored quarks give a "white" or, more precisely, colorless baryon. If the complementary colors are assigned to antiquarks—the antired is green, the antiyellow is violet, and the antiblue is orange, then the colorless antiquark will be a meson—the combination of a colored quark with the corresponding anticolored antiquark.

Each of the two degrees of freedom is associated with its own symmetry group. It turns out that colored degrees of freedom participate in strong interactions while flavors are associated with weak interactions.

It is still too early to talk about an analogy with electrodynamics. How can one describe quark interactions? The interaction between colored quarks is defined by an additional gauge field, the Yang–Mills field. This field is distinguished formally from that introduced earlier: The symmetry group of colored states is larger than the isospin group.

The symmetry group of color is the group SU(3), the group of special unitary

3×3 matrices.[2] The group SU(3) has dimension 8. From general considerations it follows that there must be eight components in the gauge field. This field is called a gluon field precisely because it glues the quarks together. The properties of a gluon field are very unusual. It is massless and carries a color charge.

If we return to real physics and try to determine how all these constructions correspond to reality, complicated problems arise that are still unresolved. For example, the fact that quarks have not been observed at any attainable energy forces one to suppose that they cannot exist as free particles. It has been suggested that strongly interacting quarks can "escape" from hadrons only as colorless groups. This proposed hypothesis has been called the problem of *confinement*. The gluon field ought to be the "mortar" that keeps the quarks inside.

The equations of quantum chromodynamics possess one remarkable feature that encourages optimism. At small distances, quarks that interact by means of of Yang–Mills gluon fields behave like free point particles. This feature is inherent in all Yang–Mills equations with a non-Abelian (noncommutative) gauge group and is called *asymptotic freedom*.

At large distances the interaction grows without bound and, one would think, ought to keep the quarks from escaping. Another vexed question is associated with the observation of a gluon field. In August 1979 a communication from the DESY accelerator in Germany reported observing the disintegration of an electron-positron pair into hadrons, which can be interpreted as the manifestation of a gluon field. These results were later repeatedly confirmed and now form the basis of the modern picture of the nature of elementary particles.

Let us leave the theory of strong interactions and move to the theory of weak and electromagnetic interactions. The phenomena associated with weak interaction are the same age as atomic physics. The first and best-studied process, the reaction of β-decay, was discovered at the turn of the present century. The process of β-decay is also called β-radioactivity because it is characteristic of radioactive nuclei. In a typical β-decay reaction in a nucleus, a neutral particle—a neutron—spontaneously decays into a proton and an electron (a β-particle). The positively charged proton remains in the nucleus; the charge of the atom in this case changes by one.

The nature of the process of β-decay was not understood until many years after the first experiments and is associated with dramatic events. The β-decay reaction threatened one of the most stable laws of physics—the law of conservation of energy. Suppose that the β-decay reaction proceeds according to the scheme

$$n \mapsto p + e^-.$$

The energy of neutron decay, that is, the energy of a proton and electron, is significantly less than the energy of the neutron. This also violates the proper connection of spin in the steady state because spin and angular momentum are not preserved. For

[2]The group SU(3) consists of complex matrices A of the order 3 with the additional relations $AA^* = A^*A = I$, $A = (a_{ij})$ and $A^* = (\bar{a}_{ji})$, where the bar indicates the operation of complex conjugation and I is the identity matrix.

example, a neutron, a particle with spin of 1/2, decays into two particles with a total spin of 1.

Physicists found themselves with a complex problem. In discussing the problem of β-decay, the famous scientist Peter Debye said, "It is better not to think about it at all—like new taxes." The deliberations of a group of the most prominent physicists in Tübingen in 1930 were interrupted by a telegram from Wolfgang Pauli, who proposed a wonderful solution to the problem: There must be yet another particle. This particle, which Enrico Fermi later dubbed a neutrino, possesses very unusual properties: it has rest mass equal to 0, it moves with the speed of light, with spin of 1/2, and is electrically neutral.

Pauli's brilliant guesswork is especially remarkable if one remembers that in 1930 there were only two elementary particles: the electron and the proton. It was another two years until the neutron was discovered by James Chadwick, and 26 years would pass before the discovery of the neutrino itself.[3] Even so prominent an authority as Niels Bohr leaned toward the idea that conservation laws might possibly be violated in elementary processes in the nucleus.

Using the neutrino hypothesis, Fermi constructed a theory of β-decay, the first theory of weak interactions. β-decay is viewed as the reaction $n \mapsto p + e^+ + \tilde{\nu}$ where $\tilde{\nu}$ is a neutrino, or more precisely, an antineutrino, a particle that differs from a neutrino only in one quantum number (the lepton charge). Fermi's theory is based on "form and likeness," as is quantum electrodynamics. In his theory Fermi proposed that direct (contact) interaction between particles occurs without the exchange of any sort of auxiliary particles. If we denote the wave functions of these particles by ψ_e, ψ_n, $\bar{\psi}_p$, and $\bar{\psi}_\nu$, then the transition amplitude in β-decay is described as $g(\psi_j \bar{\psi}_p \psi_e \bar{\psi}_\nu)$, where g is the four-Fermi coupling constant. The Fermi constant is the dimensional constant $g \simeq 10^{-5} m_p^{-2}$, where m_p is the mass of a proton. In the usual units, $g = 10^{-5} \hbar^3 c^3 m_p^{-2} \sim 10^{-5} \mathrm{GeV}^{-2} = 10^{-49}\,\mathrm{erg} \cdot \mathrm{cm}^3$. One would think that such a small amount of coupling would bring joy to everyone. Recall that the success of calculations in quantum electrodynamics is due to the smallness of the fine structure constant—1/137. Unfortunately, in the theory of weak interactions, the situation is fundamentally different.

Expansion into a series using perturbation theory makes sense only given a dimensionless parameter. Here such a quantity is $s = gE^2$, where E is energy. Therefore this parameter increases as the energy increases, and calculations based on perturbation theory become nonsensical. Nonetheless, in a majority of experiments with weak interactions, where the real energies are relatively small, estimates based on Fermi's theory lead to excellent agreement with experiment.

Theoreticians, of course, were not satisfied with such a state of affairs. They proposed to get around the difficulties caused by the discrepancies by introducing an additional particle—the carrier of interaction. This particle was supposed to perform the function that the photon performs in electrodynamics, but, considering the

[3]The neutrino was discovered in 1956 by Frederick Reines and Clyde Cowan, Jr., in experiments on the Savannah River heavy water reactor in South Carolina. Not until 40 years later did Reines receive the Nobel Prize for this discovery. Sadly, Cowan did not live to see that happy moment.

specific character of weak interactions, to possess a series of rather "strange" properties. These particles are called W-mesons (or W-bosons). We now list the properties of W-mesons:

1. W-mesons have to be massive in order to imitate contact interaction at small energies.

2. They must have an electric charge because a transfer of charge occurs during weak interactions between particles.

3. A W-meson field must be a vector field.

To construct a theory with such unusual entities as W-mesons proved to be far from simple. The natural candidate for the role of a W-meson field—the triplet field of Yang–Mills—consists of massless particles. Because we want to include electromagnetic forces in our consideration, it is necessary to introduce massless gauge fields, that is, photons. At the same time the theory should not give rise to irremovable nonphysical infinities; as physicists say, the theory ought to be *renormalizable*.

A way out of the situation was proposed in 1964, as often happens in science, by several physicists at the same time. They called this "magical transformation" of the massless particle into a heavy particle the *Higgs effect* in honor of one of its discoverers.

The basic ideas underlying the Higgs effect are fairly clear and are associated with two concepts we have already considered. The first is spontaneous symmetry breaking; the second is the concept of a gauge field. We will take them in order. The effect of spontaneous symmetry breaking in field theory occurs also in problems of statistical physics. In essence, several systems in solid state theory correspond to special models in field theory. The example we have already seen of the Heisenberg ferromagnet appears as a model of field theory defined on a two-dimensional lattice. Let us see how the breaking of internal symmetry occurs in the simplest case of a scalar field in two-dimensional space-time. Two-dimensional space-time assumes the presence of only two coordinates: space (x) and time (t). For our purposes they are completely adequate. Let us make yet one more stipulation. We shall study classical field theory. The transition to quantum fields is a separate, very difficult question. Nonetheless, the terminology of quantum mechanics is used here since it is convenient for comparison with real theories.

As in ordinary mechanics, the equations of motion are defined by the Hamiltonian $H = T + U$, where T is kinetic energy and U is potential energy. We shall choose H to be

$$H = 1/2(\partial_t\phi)^2 + 1/2(\partial_x\phi)^2 + U(\phi). \tag{13.2}$$

The potential $U(\phi)$ is

$$U(\phi) = \frac{\mu^2}{2}\phi^2 + \frac{\lambda}{4}\phi^4. \tag{13.3}$$

Here $\lambda > 0$, but μ^2 can be negative as well as positive.

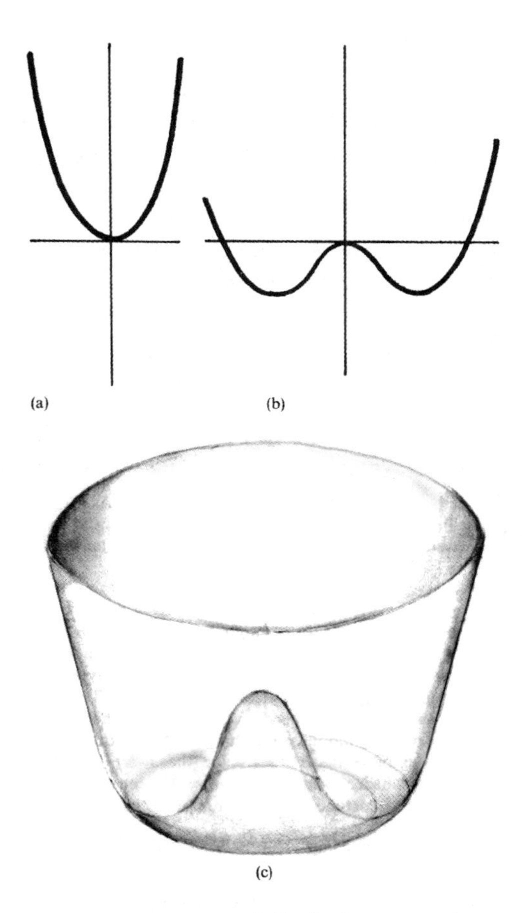

Figure 13.1: Graph of the potential $U = (\mu^2/2)\phi^2 + (\lambda/4)\phi^4$. (a) $\mu^2 > 0$. (b) $\mu^2 < 0$. (c) $U(A, B) = \lambda(A^2 + B^2 - a^2)$. The graph of the potential $U(A, B)$ is similar to the bottom of a bottle of "Napoleon" cognac. Vacua fill out the groove at the bottom and satisfy the equation $A^2 + B^2 = a^2$. The field A describes small radial perturbations of the system relative to point a while field B describes tangential perturbations (along the tangent to the vacuum groove).

The state with the smallest energy will be called the vacuum of the system and denoted $\langle\phi\rangle$. The quantity $\langle\phi\rangle$ is determined by the condition that the potential $U(\phi)$ has a minimum. The Hamiltonian (13.2) and potential (13.3) are invariants, that is, they do not change under the transformations $\phi \mapsto -\phi$.

Let us now see how the vacuums are arranged in the model. We shall draw graphs of the potential U depending on the sign of μ^2 (Fig. 13.1).

1. The quantity μ^2 is positive, $\langle\phi\rangle = 0$ (Fig. 13.1a).

2. The quantity μ^2 is negative (Fig. 13.1b).

When $\mu^2 < 0$, the minima of the potential are located at the points $\langle\varphi\rangle = \pm\sqrt{-\mu^2/\lambda}$. From the point of view of physics it is not important which of the minima is chosen as a vacuum. But no matter what minimum is chosen, the symmetry of the vacuum is spontaneously broken. The vacuum is no longer invariant relative to the symmetries of the Hamiltonian.

In quantum language particle masses are defined as the spectrum of small oscillations in a neighborhood of an equilibrium point, a vacuum. In the present case the mass of the scalar particle generated by the field ϕ (a scalar meson) is defined as the coefficient of ϕ^2 in the neighborhood of a vacuum. Thus, in the case of a system with broken discrete symmetry, the transition to a nonzero vacuum leads only to a change of sign of the mass.

A new effect appears if the equation of the field or the Hamiltonian (which is the same thing) has a continuous symmetry group. Let us look at a theory with two scalar fields A and B and the potential $U(A, B) = \lambda(A^2 + B^2 - a^2)$ (Fig. 13.1c), where λ is a fixed constant. The potential $U(A, B)$ is invariant relative to the group of rotations of the plane, SO(2)

$$A \mapsto A\cos\omega + B\sin\omega, \qquad B \mapsto -A\sin\omega + B\cos\omega. \tag{13.4}$$

The minima of the potential lie on the circle $A^2 + B^2 = a^2$. As in the preceding case, it is not important which minimum one chooses as a vacuum. But as soon as a vacuum is fixed, the internal symmetry is spontaneously broken. We shall choose

$$\langle A\rangle = a, \qquad \langle B\rangle = 0.$$

Expressing the potential $U(A, B)$ on the circle of the vacuum by the variables $A - \langle A\rangle$ and B, it is not difficult to show that the A-meson acquires mass, while the B-meson remains massless (Fig. 13.1c). Thus, given spontaneous breaking of the group SO(2), a massless meson appears. This result is general and does not depend on the particular choice of a vacuum. Only the presence of a continuous symmetry group is important. In the given case, this is the group of rotations of the plane (the circle)—SO(2).

Massless particles which appear under spontaneous breaking of continuous symmetry are called Goldstone bosons, in honor of the American physicist Jeffrey Goldstone.

In its own time the Goldstone result, proved at such a high level of rigor that it deserved the title "theorem," aroused serious concern among physicists. The presence of unusual massless particles, not observed in the real world, cast doubt on all field theories, including the mechanism of spontaneous symmetry breaking.

But as often happens with rigorous theorems in physics, the more serious the conclusions which follow from proven assertions, the more carefully one must examine the initial premises. Thus it happened with the Goldstone theory. As is said, "Without misfortune there would be no happiness." At nearly the same time another theory, the Yang–Mills theory of gauge fields, experienced similar difficulties. The Yang–Mills gauge field should have generated massless gauge vector particles. In one of his lectures, the American physicist S. Coleman characterized the resulting situation as follows:

> Now one smiles when remembering that at the time of their creations both theories—the theory of non-Abelian gauge fields and the theory of spontaneous symmetry breaking—were considered intriguing in a theoretical sense but inconsequential in a physical sense because they both predicted massless particles—gauge mesons and Goldstone bosons. Only much later was it shown that one of these illnesses is the cure for the other.[4]

The "cure" for both illnesses proved to be the Higgs mechanism. The action of the Higgs mechanism shows up in the technically simplest case: that is, the interaction of the Goldstone field ϕ with the internal symmetry group SO(2) and the gauge electromagnetic field A_μ. A detailed presentation would require pages of computations; therefore, we shall have to confine ourselves to a brief account of the effect.

We write the familiar potential $U(A, B)$ of the Goldstone field ϕ in polar coordinates:

$$A = \rho \cos\theta, \quad B = \rho \sin\theta.$$

Then the rotations (13.4) become

$$\rho \mapsto \rho, \quad \theta \mapsto \theta + \omega.$$

In polar coordinates the invariance of the potential relative to rotations signifies that U is independent of θ. Vacuums, that is, minima of the potential, lie on the circumference $\rho = q$. Fixing a point on the circle, for example $\langle\rho\rangle = a$, $\theta = 0$) we thereby choose a particular vacuum. A vacuum is not symmetric relative to the symmetry groups SO(2). A consequence of such asymmetry is the occurrence of a massless particle corresponding to a component of the θ field.

The electromagnetic field A_μ also possesses a symmetry group consisting of transformations

$$A_\mu \mapsto A_\mu - \partial_\mu\theta(x),$$

where $\theta(x)$ is an arbitrary function depending on the point x of space-time. The gauge group of transformations of the field A_μ is also isomorphic to the group SO(2).

When the Goldstone field ϕ interacts with the gauge field, a miracle occurs: the massless boson θ disappears, but in its place appears a vector particle with mass; and of course, the massive meson, corresponding to the scalar field, is preserved. The appearance of mass in an initially massless particle is called the Higgs effect.

We shall try to understand the miraculous transformation of the massless boson. Initially it was a photon—a particle with "spin" 1 and two degrees of freedom (an

[4] S. Coleman, "Secret Symmetry," in: Zichichi, A. (ed.) *Laws of Hadronic Matter*, New York: Academic Press, 1975.

electromagnetic field, as is well known, propagates in the form of transverse waves) and two scalar bosons ρ and θ. After the interaction of the photon field γ with the field ϕ the two degrees of freedom of the massless field A_μ and one degree of freedom of a Goldstone field θ combined to form three degrees of freedom of a new mass field ψ. The vector meson absorbed the Goldstone boson and acquired mass.

The true cause of the disappearance of the Goldstone boson is associated with properties of gauge invariance. The equations of the theory are invariant relative to the transformation $\theta \mapsto \theta + \omega(x)$, where $\omega(x)$ is an arbitrary function, depending on point x, the space-time coordinate. In particular, one can choose a function $\omega(x)$ equal to "minus θ," that is, by the choice of a gauge one can annihilate the θ field. Real physical quantities ought not to depend on the gauge. The destruction of the Goldstone boson means that it actually never existed. As Coleman aptly commented, "This is only a sort of gauge ghost, that is, the object, like a longitudinal photon, can be eliminated by the choice of a gauge."[5]

Very similar causes explain the interesting phenomena of superconductivity. In particular, the breaking of gauge symmetry in the macroscopic equation for superconductors in the presence of a magnetic field occurs in the unusual Meissner effect: the expulsion of a magnetic flux from superconductors. In contemporary physics a deep similarity is observed between methods developed in field theory and in the theory of condensed matter. We shall confine ourselves to a mere statement of this fact and return to the "unified" theories of weak and electromagnetic interaction. Let us see how the ideas of gauge invariance and the Higgs mechanism help resolve the difficulties that arise in Fermi's theory of weak interaction.

The most successful model of leptons was proposed in 1967 by the American theoretician Steven Weinberg and independently by Abdus Salam (1926–1996) from the International Center for Theoretical Physics in Trieste. The original version of the construction of Weinberg–Salam included only three leptons (the electron e^-, the positron e^+, and the neutrino ν) and the photon γ. In broad outline the interaction proceeds as follows.

Suppose at the outset that all particles e^-, e^+, ν are massless. Let us unite e^- and ν in one two-dimensional representation (a doublet) ψ_1 and e^+ in another (a singlet) ψ_2. The lepton field $\psi = (\psi_1, \psi_2)$ interacts with the gauge field of a photon: γ–A_μ, but this field cannot cause the field to acquire mass. Therefore the Yang–Mills field was introduced into the model. The Higgs mechanism should have led to the appearance of mass in all components of the ψ field. At this point a new difficulty arose. A neutrino is a massless particle. Consequently, it is necessary to find a way to keep one component of the field (ψ_1) massless. This introduces additional complications into the theory. It turned out that the Lagrangian gauge group of weak and electromagnetic interactions ought to be the product of two groups: groups corresponding to three lepton particles and groups preserving yet another quantum size—the lepton hypercharge.

The full group has the form $U(2) \cong SU(2) \times U(1)$. $U(1)$ is already familiar as

[5] *Ibid.*

the group of rotations of the circle, while the group SU(2) is topologically equivalent to a three-dimensional sphere S^3. This change in the theory allowed the ends to come together. Spontaneous symmetry breaking, not of the whole group U(2), but only of the subgroup SU(2), keeps the neutrino massless.

Yet another interesting peculiarity of the Weinberg–Salam model is the appearance of massive Yang–Mills gauge fields—massive vector mesons. The neutral Z-mesons and charged W-mesons should be examples; to be specific, the interactions could be mediated by the massive W, Z fields. In the model several estimates of the mass of Z- and W- mesons were obtained, showing that they ought to be very heavy, on the order of 80–100 GeV.

The possibility of generating particles of such large mass arose in the early 1980's, when a new generation of accelerators came on line. The brilliant ideas of the Soviet physicist Gersh Budker (1918–1977) (accelerators on opposed beams) and the Swiss Simon van der Meer (stochastic cooling of a beam of particles) made it possible to build accelerators with center-of-mass particle energies larger than 500 GeV. A proton accelerator on opposed beams—the $p\bar{p}$-collider—with an energy of 540 GeV was placed in operation at CERN in 1978, and by late 1982 the first group of experimenters (UA1), headed by Carl Rubia, announced the detection of W-bosons. In early May of 1983 the UA1 group detected a Z-boson. The measured masses of the W- and Z-bosons were in excellent agreement with the predictions of the theory of the Weinberg–Glashow–Salam electroweak interaction. The experiments to detect heavy bosons are a triumph of technical resources, experimental skill, and theoretical predictions, and were instantly recognized by the scientific community. The 1984 Nobel Prize in physics went to Rubia and van der Meer. More precise measurements of the masses of the W- and Z-bosons were later carried out at SLAC (Stanford) on the e^+e^--collider and at CERN (LEP1) and Tevatron in the Fermilab (Batavia, Illinois). The results of experiments confirmed the theory of electroweak interaction within 0.1%. The launching of a new series of accelerators in CERN in 1996 (LEP2) is making it possible to obtain even more precise information on the masses of heavy bosons. In the meantime, only one prediction of the theory remains unconfirmed by experiment—the existence of heavy scalar bosons—Higgs particles. A lower bound of $\sim$ 65 GeV has been given for their mass. Experiments to detect them will soon be carried out at CERN on the giant LHC (large hadron collider) supercollider, which it is anticipated will be built early in the next century (2005).

The presence of heavy mesons is a necessary consequence of the correctness of the theory. However, the theory also predicted other processes connected with the exchange of neutral Z-mesons. Such reactions, called *processes with neutral currents*, were discovered in 1972 by researchers at several scientific centers, including CERN. The discovery of neutral currents appeared in a convincing experiment which confirmed the basic propositions of the gauge theory of weak and electromagnetic interactions.

Although for a theoretician the internal beauty and harmony of a theory probably plays a decisive role, the Nobel Committee uses other criteria. The awarding of the Nobel Prize in 1979 to S. Weinberg, Abdus Salam, and S. Glashow shows that this

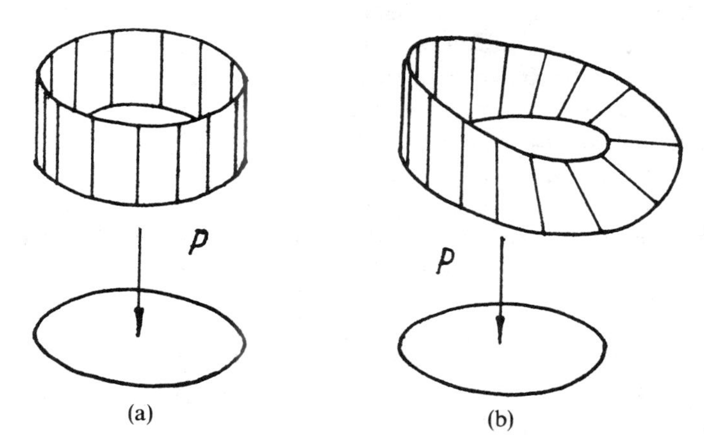

Figure 13.2: Trivial and nontrivial fiber bundles. (a) A cylinder. (b) A Möbius strip. (c) See Fig. 9.8a. The tangent fiber bundle to the sphere. The tangent fiber bundle of the sphere is nontrivial. This assertion, together with the more general one asserting the nontriviality of the fiber bundle on any two-dimensional surface which is not a torus, follows from Poincaré's hedgehog theorem.

theory ought to be included in the "golden background of physics."

Having now acquired some idea of problems facing physicists in elementary particle theory, the reader may justly ask: While all of this is very interesting, what is the relevance of topology here? To such a question one can give a one-sentence reply: Topology makes it possible to explain the unusually complex structure of solutions of equations of gauge fields. A more elaborate answer requires a small mathematical digression.

From a topological point of view, gauge fields are a special case of fiber bundles. What are fiber bundles? Consider the now-familiar Möbius strip and the cylinder. Remember that these surfaces are not homeomorphic. However, if one chooses a small neighborhood of an arbitrary point on the cylinder and the Möbius strip, it is easy to see that they have the same local structure. Such a neighborhood can be represented by combining pairs of points x, y, where $x \in S^1$, $y \in I$. Here S^1 is the circle, while I is a segment (see Fig. 13.2).

If the point x is made to traverse the whole circle S^1 and the segment I is a function of the point x, then in the case of a cylinder it turns out that segment I, when it returns to the initial point, preserves the orientation. But in the case of a Möbius strip it changes it to the opposite (rotates it by 180°). Thus, although the cylinder and Möbius strip have the same local structure, globally they are different. In this case topologists say that the cylinder is trivial, while the Möbius strip is a nontrivial fibration. Another example of a fibration which we have seen is the set of tangent vectors to a two-dimensional oriented surface. In general a fiber bundle can be represented as follows. Let X be an arbitrary space (the base) at each point of which a

Figure 13.3: The "Fair" of physical ideas. Caricature by the physicist Alvara De Rujula. Who will succeed in rescuing the quark damsel confined in her tower? (De Rujula's view of the current scene in particle theory) (CERN *Courier* No. 7, Vol. 19, Oct. 1979), reprinted with permission. Now, twenty years later, there is not much that could be changed in De Rujula's picture. The quark damsel is still in the dark, despite new discoveries and attempts to reactivate the old devices. All the old hopes, such as the modest CP^n tadpole, have been dashed, their place taken by new favorites such as the Witten–Seiberg invariants. But the author lacks the skill and imagination to describe them.

copy of the space Y is attached (the fiber). Given a point x moving along S, the fibers Y_x are transformed into each other by the action of the group G of transformations of the fiber Y. The fiber bundle E is the union of all the fibers Y_x.

From the point of view of topology gauge fields are fiber bundles in which the base is four-dimensional space-time, while gauge groups are groups of transformations of the fiber. It follows from the property of local gauge invariance that these are nontrivial fibrations.

Far-reaching analogies between topological and physical properties of gauge fields are being observed. Gauge fields are not simply arbitrary fiber bundles; they are endowed with an additional geometrical structure which permits comparison of different fibers. Such a structure is called a *connection*.

In terms of a connection one can give a geometric interpretation to the physical quantities inherent in gauge fields. In essence gauge fields and fiber bundles have the same relationship to each other that gravitational fields have to Riemannian spaces.

At present the Yang–Mills fields seem a reliable foundation for constructing a theory of elementary particles. The efforts of theoreticians are directed at solving the equations of gauge fields. The task has turned out to be extraordinarily complicated. Ingenious physicists are assaulting the Yang–Mills equations from all sides (Fig. 13.3). Various approximate schemes—refined variants of perturbation theory—have been proposed, and simpler models, possessing characteristic features of the Yang–Mills equations, are being investigated, but there is not yet a final solution.

Topological methods have proved to be a major help in solving the problems that arise. Topology makes it possible to explain the general structure of the set of solutions without even knowing their analytic expression. The most interesting result of topological investigations is the appearance of new conserved quantum quantities, having a purely topological, rather than a dynamic, origin. Such quantities are called topological charges. Solutions that carry topological charges—topological particles—have the same direct relationship to reality. The origin of such particles leads to important physical consequences. One of the most interesting examples of such a type of particle is the 't Hooft–Polyakov monopole in the Georgi–Glashow model of weak interactions, which will be the subject of a separate discussion.

Chapter 14

Topological Particles

IN 1974 the Dutch physicist Gerardt 't Hooft and the Russian physicist Alexander M. Polyakov (now at Princeton University) found solutions to the Yang–Mills equation for the group SO(3) and additional scalar fields (the fields of Goldstone–Higgs).[1] These solutions had one topological charge which was interpreted as a "magnetic charge." G. 't Hooft suggested calling them magnetic monopoles. The name was to symbolize the deep commonality of the newly discovered "particle" with another mysterious object—the Dirac magnetic monopole.

Let us go back fifty years and to Dirac's paper, "Quantum singularities in an electromagnetic field," published in 1931. In this paper Dirac gave a completely new interpretation of two questions that are fundamental to physics. Why are there particles carrying an electric charge in nature but none with a magnetic charge? In suitable units the charges of all particles turn out to be integer multiples of the charge of an electron e. Dirac suggested that magnetic monopoles exist and found that this hypothesis leads to a natural explanation of the quantization of electric charge. By this elegant analysis he showed that the presence of a magnetic monopole does not lead to any theoretical contradictions with contemporary physical concepts, in particular, with the Maxwell equations. In his characteristic style, he concluded: "Under these circumstances one would be surprised if Nature had made no use of it."

Dirac monopoles must possess a number of surprising properties. Just as for electrically charged particles, they must also fulfill the condition of conservation of magnetic charge. This means that a monopole, once created, cannot by itself disappear without having collided with another monopole with a magnetic charge of the opposite sign. The elementary magnetic charge of a monopole must be 137/2 times the charge of an electron. Consequently, the force of interaction of two monopoles must exceed the force of the interaction of two electrons at the same distance by approximately a factor of 4692.

"This very large force may perhaps account why poles of opposite sign have never

[1] Before the discovery of neutral currents, the Georgi–Glashow model of weak interactions competed with the Weinberg–Salam model.

yet been separated," wrote Dirac in 1931.[2]

Sixty years of effort to detect Dirac monopoles have not changed this state of affairs. So far, there are no decisive arguments proving the reality of magnetic monopoles, but one cannot exclude their future appearance. Nonetheless, there is serious doubt that they exist. The introduction of the concept of gauge invariance into the theory of elementary particles has provided a completely new look at the problem of monopoles. A decisive step in this direction was taken in the work of 't Hooft and Polyakov. They succeeded in constructing a steady-state solution (that is, one independent of time) with finite energy in the Georgi–Glashow model. They interpreted such a solution as a particle; and having computed its mass, they showed that it agrees with estimates of the mass of a charged W-boson. Another interesting property of this solution is the appearance of a "magnetic" charge.

Spontaneous symmetry breaking, the Higgs mechanism, nontrivial fibrations—all themes of our previous chapters—play their part in the concept known as a monopole.

Let us consider in somewhat greater detail the 't Hooft–Polyakov construction. We shall not write out the Lagrangian of the Georgi–Glashow model and the corresponding equations of motion; instead we shall formulate only those properties of the Yang–Mills and Higgs fields used in finding solutions. Their physical interpretation was discussed in the chapter on gauge fields. Two kinds of fields are involved: the Yang–Mills vector fields W^i_μ and the Higgs scalar field ϕ^i, which assumes values in isotopic space—the three dimensional space R^3. The gauge group SO(3) is the group of rotations of R^3. The spatial index μ assumes the four values 0, 1, 2, 3, and the isotopic index i assumes the three values 1, 2, 3. Because we shall be considering only stationary solutions, we shall assume that the time component (t) is 0. The Higgs potential $U(\phi)$ is invariant relative to the group SO(3) and is equal to

$$U(\phi) = \frac{\mu^2}{2}(\phi \cdot \phi) + \frac{\lambda}{4}(\phi \cdot \phi)^2,$$

where

$$(\phi \cdot \phi) = (\phi^1)^2 + (\phi^2)^2 + (\phi^3)^2. \tag{14.1}$$

The set of minima of the potential $U(\phi)$ defines the vacuum space Ω and satisfies the relationship

$$\frac{\partial U}{\partial \phi} = 0; \qquad \frac{\partial^2 U}{\partial \phi^2} \geq 0.$$

In this case the space Ω is the two-dimensional sphere S^2.

One can fix a vacuum by choosing a certain point on the two-dimensional sphere. The simplest vacuum solution, which is constant in space, will be called the Higgs vacuum. It has the following form:

$$\phi^i(x) = \frac{\mu}{\sqrt{2}} e_3, \qquad W^i_\mu(x) = 0. \tag{14.2}$$

Here e_3 is a unit vector in the space $\mathbb{R}^3$, directed along the z axis.

[2] P.A.M. Dirac, *Proc. Roy. Soc.*, A. **133** (1931), p. 60.

The other class of vacuum solutions was found by 't Hooft and Polyakov. They showed that there exist solutions to the Yang–Mills equations which do not depend on time and have the following form as $|x| \to \infty$:

$$\phi^i(x) = \frac{x^i}{|x|}, \quad W^i_\mu(x) = \varepsilon_{\mu ij}\frac{x^i}{|x|^2}. \tag{14.3}$$

Here $\varepsilon_{\mu ij}$ is an antisymmetric tensor. For $\mu = 1, i = 2, j = 3$ it is given by $\varepsilon_{\mu ij} = 1$ for even permutations of the subscripts and $\varepsilon_{\mu ij} = -1$, for odd permutations, because the time component ($\mu = 0$) equals 0. The happy term hedgehog was invented for such a solution. The spines of the hedgehog form a radial vector field.

The nontriviality of this solution consists of the fact that for given finite values (x) no closed-form solution is known. The existence of solutions with such asymptotic properties and the even more general properties

$$\phi^i(\mathbf{x}) \sim \phi(\mathbf{n}), \quad W^i_\mu(\mathbf{x}) \sim \frac{1}{r}W^i_\mu(\mathbf{n}), \quad |\mathbf{x}| \to \infty, \quad (\mathbf{n}\cdot\mathbf{n}) = 1,$$

appears as a purely topological fact.

Very little analysis is required to explain why a solution of the form (14.3) is interpreted as a monopole. Let us go back a little and look carefully at the Higgs vacuum (14.2). The Higgs field ϕ^i is directed along the z-axis and therefore remains invariant under rotations around the z-axis.

The Higgs mechanism remains a massless particle, connected with the group of rotations about the z-axis. The corresponding massless particle is identical to the photon γ. In precise analogy with the theory of electromagnetism, one can connect the electromagnetic field $F_{\mu\nu}$ and write the Maxwell equations for it in terms of the fields ϕ^i and W^i_μ. The field $F_{\mu\nu}$ occurs as a physically observable quantity and must be independent of the choice of the gauge. Therefore it is defined for any vacuum solution, in particular, for a hedgehog. Direct computation of the field $F_{\mu\nu}$ led to startling results:

$$F_{\mu\nu} = \frac{-1}{er^3}\varepsilon_{\mu\nu i}r^i.$$

Up to sign this $F_{\mu\nu}$ corresponds to the radial magnetic field B of a point source with magnetic charge g

$$g = \frac{4\pi}{e}, \quad B = \frac{r^a}{er^3}. \tag{14.4}$$

The quantity e is associated with the electrical charge $\tilde{e}$ by the relationship $\tilde{e} = e\hbar/2$. The 't Hooft–Polyakov solution thus acquired all necessary properties to merit being called a magnetic monopole.

Let us now discuss monopole solutions from a topological point of view. It is intuitively clear that the vacuum solution (14.2) and the hedgehog (14.3) are topologically different, but how can we prove this? The techniques of homotopy theory again come to our aid. The solution determined using the functions ϕ^i amounts to nothing more than a mapping of the coordinate space $\mathbb{R}^3$ with certain boundary conditions into the vacuum isotopic space Ω. The asymptotic conditions give a mapping of the unit sphere S^2 into the vacuum space Ω. In order to avoid confusion, we shall write the "vacuum" sphere with a tilde—$\widetilde{S}^2$.

As noted above, the set of topologically inequivalent mappings of the sphere S^2 into the sphere $\widetilde{S}^2$ is called the second homotopy group $\pi_2(\widetilde{S}^2) = \mathbb{Z}$, where $\mathbb{Z}$ is the group of integers and where the group operation is addition.

It is now clear how to prove the inequivalence of the two vacuum solutions. It suffices to show that they belong to different homotopy classes. But this is already quite obvious. The vacuum solution (14.2) maps the sphere S^2 to the point $x_0 \in \widetilde{S}^2$; the hedgehog maps the sphere S^2 onto the sphere $\widetilde{S}^2$ identically. From this it follows that one cannot transform the solution (14.3) into (14.2) by a continuous transformation.

One of the most interesting consequences of the foregoing discussion is a simple explanation of the appearance of a "Dirac string." Recall that the original Dirac construction required a line singularity emanating from a monopole and on which the vector potential A_μ undergoes a break. The requirement that observable physical quantities (in particular, a magnetic field) be finite led to special quantization rules when traversing the string. In this connection a question arose as to the observability of the string. In the 't Hooft–Polyakov monopole this problem does not exist. The origin of the line of singularity has a very simple explanation. The transition from a hedgehog vacuum to a Higgs vacuum is possible only through a broken gauge transformation, since they belong to different homotopy classes. The lines of the break of the vector potential will be "Dirac strings." It is also obvious that a Dirac string can have a completely arbitrary form and direction in space.

From topological considerations it also follows that there is an infinite set of vacuums, parametrized by the integer k. One could call them "twisted hedgehogs," where the usual hedgehog corresponds to the value $k = 1$. The magnetic flux defined by a k-twisted hedgehog, would be k times larger than for a normal hedgehog.

The appearance of monopoles in the Yang–Mills equations has made possible a fresh look at the Dirac monopole. The 't Hooft–Polyakov monopole is a classical regular solution, appearing in a whole class of models of weak and electromagnetic interactions. Such a solution is absent in the original scheme of Weinberg–Salam but appears in several of its modifications.

The purely topological origin of monopole solutions became clear immediately after the work of 't Hooft and Polyakov. A criterion was obtained for the existence of solutions of monopole type in the Yang–Mills equations with an arbitrary symmetry group in terms of homotopy groups. Do magnetic monopoles then really exist? Concluding our brief discussion of this most interesting theme, I would like to talk about the somewhat unexpected possible "observation" of a monopole. It has already been mentioned that monopoles must be very massive particles in all gauge field theories. The energies obtained in contemporary accelerators are so far inadequate for creating particles with such masses. But there are situations in which monopoles have been observed already, or at least may have been observed. The mysterious boojums that occur on the surfaces of cholesteric liquid crystals or ^{3}He-A might be indicators of monopoles. The role of monopoles is filled by a vortex located in the center of the volume, relaxing onto a special point of the surface, the boojum.

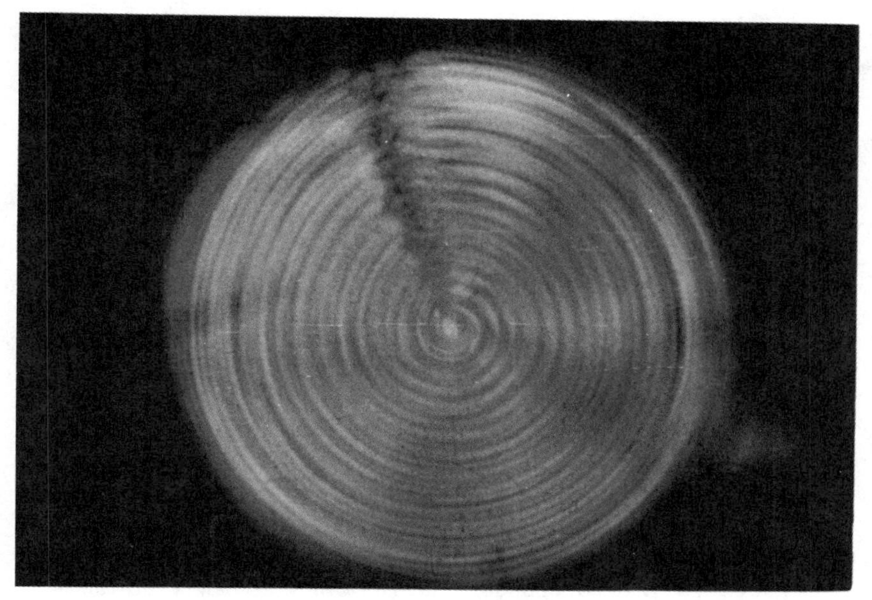

Figure 14.1: Monopole in a cholesteric drop. *Courtesy O. Lavrentovich.*

Theoretically, there is nothing to prevent such a possibility. In any case, the topological nature of the appearance of monopoles in gauge theory and of a field of vortices in ordered media is one and the same. This is beautifully shown in the experiment conducted by Oleg Lavrentovich (Kent State University), pictures of which are shown in Figs. 14.1 and 14.2.

In addition to the topological methods already familiar to us from solid-state physics—homotopy and homology theories—in field theory one has to use the full power of modern mathematics. By itself topology is completely inadequate for solving the Yang–Mills equations.

To see the truth of these words let us turn to yet another class of nontrivial topological solutions of the Yang–Mills equations, one that was discovered soon after the gauge monopoles and has played an extremely important part in the whole subsequent history of the interaction of topology and physics.

14.1 Instantons, or Pseudo-Particles

A remarkable property of the current stage of development of physics is the intimate intertwining of the ideas and methods of field theory and condensed matter theory. Many model systems in field theory correspond to real objects in solid state physics, and a number of concepts that originally arose in statistical physics are now finding

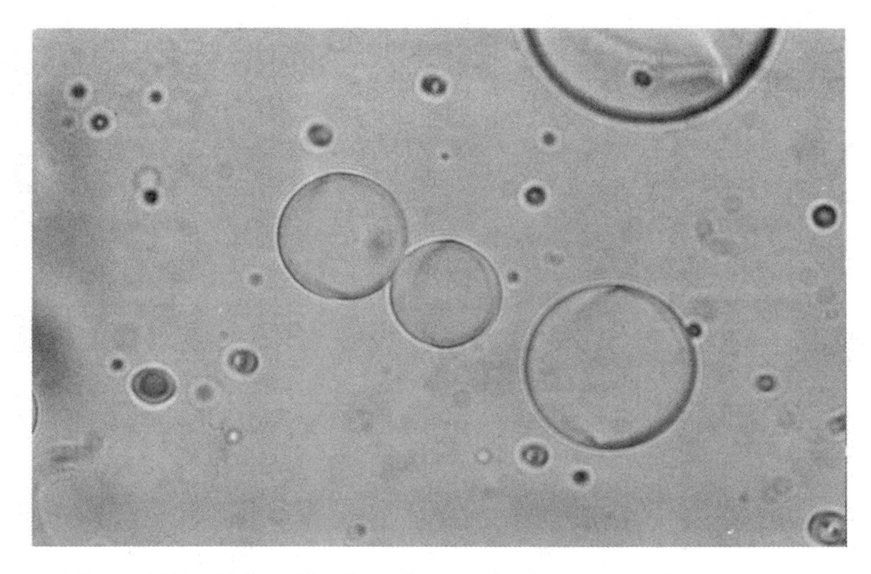

Figure 14.2: Pairs of boojums in nematic drops. *Courtesy O. Lavrentovich.*

important applications in field theory. This unity of modern physical theory is quite remarkable. It is no wonder that a number of theoreticians are working successfully in both areas of physics.

Two problems, one in the elementary particle physics, the other in statistical physics, are of fundamental significance for the development of the theory. In elementary particle theory the problem is that of "confinement" of quarks; in statistical physics it is the theory of phase transitions.

In slightly more detail, it is a question in the first instance of constructing a coherent theory based on the Yang–Mills equations to explain the confinement of quarks; in the second case it is a matter of describing phase transitions in the vicinity of a critical point (a λ-point). These fundamental problems arise in connection with the study of the fluctuations of a vacuum (field theory) and the ground state (critical phenomena).

The Lagrangians that define the corresponding theories are essentially nonlinear, so that finding solutions of the equations they generate is a difficult mathematical problem. How do physicists, who think pragmatically, proceed in such cases? They attempt to find a simpler solvable model that retains the properties of realistic systems.

An important, yet at the same time technically rather simple, model is the well-known two-dimensional Heisenberg ferromagnet (see Chapter 11). It was the continuous analogue of this model that A. Polyakov (partly in collaboration with his colleague from the Landau Institute A. Belavin) used as the starting point for a series of papers on critical phenomena in two-dimensional systems, which led to the discovery of instantons.

Let us denote by $n^a(\mathbf{x})$ the spin vector that assumes values in the sphere S^2, that is, $(n^a)^2 = 1$, where the superscript a assumes the values $a = 1, 2, 3, \ldots$. The action of S has the form

$$S = \int (\partial_\mu n^a)(\partial_\mu n^a)\, d^2\mathbf{x} \tag{14.5}$$

where $d^2\mathbf{x}$ is the element of area on the plane $\mathbb{R}^2$ and $\mu = 1, 2$. We shall be interested in solutions of the equation

$$\delta S = 0 \tag{14.6}$$

with finite action. We shall explain in brief outline the importance of such solutions for problems of phase transitions in the model (14.5). For the reader interested in the physics of this phenomenon I recommend the books [Po] and [PP], where the current state of the subject is presented fully and accessibly.

The principal object in terms of which the type of phase transition is determined is the spin correlation function: $\langle \mathbf{n}(\mathbf{x})\mathbf{n}(\mathbf{x}_1)\rangle$. Here $\langle\cdot\cdot\rangle$ denotes the average over all possible configurations with the weight function $\exp(-S/k\mathrm{T})$, where k is the Boltzmann constant and T the temperature. All the properties of the system are determined by analyzing the statistical sum (partition function):

$$Z = \sum_{\text{over all configurations}} e^{-S/k\mathrm{T}}.$$

As $\mathrm{T} \to 0$ the local action minima S, that is, the solutions of Eq. (14.6), begin to play an essential role in the behavior of the correlation functions.

Besides the trivial solution $\mathbf{n}_0(\mathbf{x}) = \text{const}$, Eq. (14.6) may have other local momenta. They play a fundamental role in the possible change in the nature of the phase transition. The corresponding arguments involve an analysis of the symmetry of Eq. (14.6). The solutions of Eq. (14.6) with finite action have conformal symmetry, in particular gauge invariance. These solutions were called *pseudo-particles* or *instantons*.

It follows from the gauge invariance of the solutions that the mean distance between instantons at small T is of the same order as their dimensions, that is, instantons can be thought of as drops in a homogeneous medium whose distances are of the same order as the dimensions of the pseudo-particles themselves. Such random inhomogeneities violate spin correlation at a distance $R > r$ and thereby change the nature of the phase transition. In the present case this indicates the absence of a phase transition.

All this reasoning is supported on a firm topological foundation, to whose analysis we now turn. How can we find all solutions of Eq. (14.6) with finite action? Consider the class of solutions that have the following boundary behavior:

$$\mathbf{n}(\mathbf{x}) \to \mathbf{n}_0 \text{ as } |\mathbf{x}| \to \infty. \tag{14.7}$$

Then the spin variable $\mathbf{n}(\mathbf{x})$ is defined on the extended plane $\mathbb{R}^2 \cup \infty = S^2$. Since the metric on $\mathbb{R}^2$ is conformally equivalent to the metric on S^2, all solutions of Eq.

(14.6) satisfying the condition (14.7) have a unique extension to S^2. Thus we have a mapping

$$\mathbf{n} : S^2_{\text{space}} \to S^2_{\text{spin}}. \tag{14.8}$$

Since physical states that pass into one another by a continuous deformation are indistinguishable, we are interested in the topologically nontrivial mappings (14.8). We have already encountered a topological object that classifies the nontrivial mappings (14.7) when we analyzed the point defects in a nematic and when we described monopoles. This object is the homotopy group $\pi_2(S^2) = \mathbb{Z}$. An integer k ($k \in \mathbb{Z}$) characterizing a mapping can be represented in integral form:

$$k = \frac{1}{8\pi^2} \int\limits_{\mathbb{R}^2} \varepsilon_{\mu\nu}\mathbf{n} \cdot (\partial_\mu \mathbf{n} \times \partial_\nu \mathbf{n})\, d^2\mathbf{x}. \tag{14.9}$$

Here $\varepsilon_{\mu\nu}$ is a well-known antisymmetric tensor: $\varepsilon_{12} = 1$, $\varepsilon_{21} = -1$, $\cdot$ is the inner product, and $\times$ is the cross product.

The possibility of such a representation follows from classical theorems of topology. (For the exhausted reader we shall say that this is a well-known theorem of Hurewicz.) So as not to depart from the general plan of this book I am forced to confine myself to listing the facts needed for what follows. The interested reader can find the corresponding proofs and details in a number of surveys and books, for example [Pol] and [Mo]. The representation of the quantity k in the form (14.9), which is called a *topological charge*, makes it possible to prove the following key inequality:

$$S \geq 4\pi k, \tag{14.10}$$

which is equivalent to the condition

$$\partial_\mu \mathbf{n} \pm \varepsilon_{\mu\nu}\mathbf{n} \times \partial_\nu \mathbf{n} \geq 0. \tag{14.11}$$

The equality $S = 4\pi k$ means

$$\partial_\mu \mathbf{n} \pm \varepsilon_{\mu\nu}\mathbf{n} \times \partial_\nu \mathbf{n} = 0. \tag{14.12}$$

This equation is called the *duality* equation (with the positive sign) or the *antiduality* equation (with the negative sign). Solutions of Eq. (14.12) are also called *instantons* (with the positive sign) or *anti-instantons* (with the negative sign). In this model, as we shall explain in a moment, both definitions of instantons give the same class of solutions, though in more complicated situations not every solution of the equation for an extremal (Eq. (14.6)) with finite action is a superposition of solutions of the duality and antiduality equations. For that reason instantons and anti-instantons are defined in general as solutions of Eq. (14.12).

Equation (14.12) admits explicit solutions if we pass to complex coordinates after a preliminary mapping of the Riemann sphere S^2_{space} (the z-plane) using stereographic projection onto the w-plane (X^2_{spin}):

$$w = w_1 + i w_2 = \frac{n^1 + i n^2}{1 - n^3} = \cot(\theta/2)\exp(i\varphi).$$

In (z, w)-variables the expression for the action (14.9) is

$$S = 2 \int_{\mathbb{R}^2} \frac{d^2\mathbf{x}}{(1+|w|^2)^2} \Big(\frac{\partial w}{\partial z} \frac{\partial \bar{w}}{\partial \bar{z}} + \frac{\partial w}{\partial \bar{z}} \frac{\partial \bar{w}}{\partial z} \Big)$$

and the expression for the topological charge (14.10) is

$$Q = \frac{1}{\pi} \int \frac{d^2\mathbf{x}}{(1+|w|^2)^2} \Big(\frac{\partial w}{\partial z} \frac{\partial \bar{w}}{\partial \bar{z}} - \frac{\partial w}{\partial \bar{z}} \frac{\partial \bar{w}}{\partial z} \Big).$$

Hence it follows that the duality (antiduality) equations reduce to the Cauchy–Riemann equations $\frac{\partial w}{\partial \bar{z}} = 0$ (duality) or $\frac{\partial \bar{w}}{\partial z} = 0$ (antiduality).

Taking account of the boundary conditions (14.7), we obtain the general solution of Eq. (14.12) in the form of a rational function on the sphere S^2

$$w(z) = \prod_{i=1}^{k} \frac{z - a_i}{z - b_i}.$$

Here we have chosen the condition $n(\infty) = n_0 = 1$. It is obvious that such a choice of boundary condition causes no loss of generality in the solution of Eq. (14.12), since because of the invariance of the action of S with respect to the group of rotations of the sphere (the group SO(3)), any boundary value can be translated to 1.

The topological charge Q equals k for an instanton and $-k$ for an anti-instanton. The solution is characterized by $(4k - 3)$ real parameters: the $2k$ complex numbers a_i, b_i determine $4k$ parameters, but one must take account of the fact that the group SO(3), which has dimension 3, acts globally on the sphere. For that reason the total number of parameters is $4k - 3$.

The paper of Belavin and Polyakov (published in 1975) drew a large response, since it contained a clearly articulated investigation of the role of the classical solutions in the study of fluctuations in the vicinity of a critical point. From a purely mathematical point of view this paper contained no new results. Indeed, 20 years before this paper appeared the American mathematician F.B. Fuller, and subsequently J. Eells, J.H. Sampson, and a number of others had studied variational problems for action functionals significantly more general than the example investigated by Belavin and Polyakov. The corresponding class of mappings is called *harmonic*, since in the case of a mapping of a Riemannian manifold M^n (where $n = \dim M$) into $\mathbb{R}^1$ harmonic mappings are simply harmonic functions.

Although there already existed a voluminous literature on harmonic mappings at the time when the paper of Belavin and Polyakov appeared, physicists were completely unaware of it. The period of close interaction between mathematicians and physicists had only just begun, and the situation began to change overnight and led to a bold effort to study the classical solutions of a significantly more complicated system—the Yang–Mills equations. Polyakov and Belavin were able to obtain the valuable advice of a famous topologist, S. Novikov. A result of their contact with another Moscow topologist, A. Schwarz was a remarkable paper by four authors. (The fourth letter in this abbreviation stands for Yu. Tyupkin, a student of Schwarz.) In this paper an instanton solution of the Yang–Mills equation was found (having charge 1).

The work on instantons inspired a steady stream of research, which can be summarized as follows:

1. *Physical applications.* Soon after the discovery of instantons V. Gribov and G. 't Hooft proposed an intuitive and physically important interpretation of instantons as transitions between different vacua (Lagrange–Yang–Mills minima) in Euclidean space. In analogy with quantum mechanics instantons—the classical solutions of the Yang–Mills equation—tunnel between different vacua. The contribution to the corresponding vacuum transition amplitudes turns out to be exponentially small and cannot be obtained from perturbation theory. Taking account of the instanton contribution has led to profound results in quantum chromodynamics. As frequently happens in the history of science, the main problem for which instantons had been invented (confinement of quarks) still had not been solved. Taking account of instantons was not sufficient for the computation of large-scale fluctuations of a vacuum. Still, the role of instantons in particle physics was significantly greater than just their use in solving several important particular problems. Along with monopoles, instantons brought into physics a kind of mathematics that was unusual for it—topology, and later algebraic geometry as well. The connection with these branches of mathematics, which was not traditional for physics, has done much to shape the present situation in theoretical physics. The opposite influence has been of no less importance.

2. *Mathematical applications.* Originally the mathematical papers were concentrated around the analysis of the topological structure of instanton solutions. In particular the most complete results on the classification of k-instanton solutions were obtained by the methods of algebraic geometry in a paper of M. Atiyah, V. Drinfel'd, N. Hitchin, and Yu. Manin (the ADHM-solutions). But the most remarkable applications of instanton theory were to be the discovery by S. Donaldson of simply connected 4-dimensional manifolds with different smooth structures. Unfortunately any reasonably complete discussion of these results is beyond the scope of a popular exposition. The interested and sufficiently qualified reader will find a complete exposition of Donaldson's theory in the books [DK] and [FU]. I shall try to describe the key elements of Donaldson's construction and point out several corollaries that are of both mathematical and general physical interest.

3. *Donaldson's theory.* The topological classification of simply connected 4-dimensional manifolds M^4 is based on the study of the algebraic properties of the so-called *intersection* quadratic form Q defined on the 2-dimensional homology group $H^2(M^4, \mathbb{Z})$. A fundamental classification theorem due to

M. Freedman asserts that any unimodular quadratic form over the ring of integers $\mathbb{Z}$ can be the intersection form Q on $H^2(M^4, \mathbb{Z})$. For the time being we shall regard the manifold M^4 as being topological, that is, the topology of M^4 is given by its continuous functions. As often happens in mathematics, behind the simple statement of the theorem lurks an exceptionally difficult proof. The idea of classifying manifolds using the intersection form originated in 1952 in papers by the famous topologist V. Rokhlin. He obtained a result that was fully appreciated only 30 years later. He showed that if a smooth compact simply connected manifold M^4 has an even intersection form Q (that is, it assumes even values), then its signature σ must be divisible by 16, while for a topological manifold it suffices that it be divisible by 8. As Donaldson later showed, a manifold M^4 with an even form Q cannot have a smooth structure. Freedman's paper was published in 1982, and only a year later Donaldson showed that for a smooth simply connected M^4 with a positive-definite form Q the form Q can be diagonalized over $\mathbb{Z}$ (the ring of integers), that is, Q is odd.

$$Q = x_1^2 + \cdots + x_n^2. \tag{14.13}$$

Comparison with Freedman's results leads to a fundamental corollary—the existence of 4-dimensional manifolds that are homeomorphic but not diffeomorphic. It is striking that for the proof of this theorem of algebraic topology Donaldson had to invoke the theory of instantons. The idea of the proof is as follows.

Consider the manifold $\mathbb{C}P_n^2$ that is the sum of n copies of the 2-dimensional complex projective plane $\mathbb{C}P^2$. For this manifold, which satisfies the hypothesis of Donaldson's theorem, one can compute the form Q. The next step is to construct a 5-dimensional manifold V^5 with two boundary manifolds, one of which coincides with a manifold M^4 and the other with $\mathbb{C}P_n^2$. The existence of such a triple $(V^5, M^4, \mathbb{C}P_n^2)$ follows from purely topological reasons. In topology such a construction is called a *cobordism*, and the manifolds M^4 and $\mathbb{C}P_n^2$ which can be joined by the "film" V^5 are said to be *cobordant*. Since the signature σ of the quadratic form Q is invariant under cobordisms, it follows that the form can be reduced to (14.13) for an arbitrary M^4. The most difficult part of Donaldson's proof is the construction of the required cobordism. That is the stage at which it is necessary to make full use of the fact that on 4-dimensional manifolds there exist smooth bundles with the connection generated by the Yang–Mills fields, more precisely, the connection generated by the duality equations, that is, instantons.

Donaldson's result stimulated a series of papers in which the rich and completely unexpected structure of 4-dimensional manifolds was discovered. The most surprising result is probably the discovery of different smooth structures on noncompact manifolds, in particular on a space homeomorphic to $\mathbb{R}^4$. The American mathematicians R. Gompf and C. Taubes constructed respectively a countable and a continuous family (these families are not the same) of different

> smooth structures on $\mathbb{R}^4$.[3] Since Euclidean space and Minkowski space-time are the basic objects in all fundamental physical theories and the concept of a metric is closely connected with smoothness, this discovery may have deep significance for the modern theory. It suffices to mention quantum gravity and the theory of membranes and strings, where it is necessary to integrate over all metrics in order to compute the generating functionals. While physicists are overcoming difficulties of a more prosaic kind, 4-dimensional smooth topology has become an object of intensive study by mathematicians. Donaldson's theory, in particular, has been extended to nonsimply connected manifolds. One interesting result of this research was the clarification of a surprising property of smooth 4-dimensional manifolds: a number of geometric invariants of smooth manifolds are the same as those of 4-dimensional algebraic varieties, which are manifolds with a significantly more rigid structure than topological manifolds. This amazing phenomenon is characteristic of 4-dimensional topology.

Instantons were originally found by physicists as Yang–Mills connections in a bundle over the sphere $\mathbb{S}^4$. Ironically, despite the whole development of the theory, the most natural question remains unanswered: Do there exist different smooth structures on $\mathbb{S}^4$? (If so, the group $H^2(\mathbb{S}^4, \mathbb{Z})$ is zero and the existing theory is inapplicable.) The connection of Yang–Mills fields with 4-dimensional topology by no means exhausts the application of physical ideas in modern mathematics. Very interesting and promising is the connection of field theory with knot theory and three-dimensional topology. But we shall discuss that below.

Returning, after our brief journey into the world of higher theory, to the problems faced by physicists directly engaged in studying the structure of elementary particles, one must frankly confess that the basic method of obtaining numerical results that can be compared with experiment is still the method of perturbation theory.

Until recently the only means for obtaining numerical quantities in field theory was the perturbation method. This method, which has given excellent results in quantum electrodynamics, is not so effective in theories of strong interactions; but it is used for lack of anything better. One should not think that perturbation theory in the theory of gauge fields is a simple thing. The efforts of many outstanding theoreticians have resulted in the construction of an invariant theory of perturbations for the Yang–Mills fields in the framework of which the Yang–Mills equations have successfully been quantized.

In recent years, however, yet another area of investigation has arisen, on which specialists in field theory are placing high hopes: solitons.

---

[3] Spaces homeomorphic, but not diffeomorphic, to $\mathbb{R}^4$ are called "fake" $\mathbb{R}^4$'s and denoted $\mathbb{R}^4_f$. They have amazing topological properties. For example, in contrast to the standard $\mathbb{R}^4$, the spaces $\mathbb{R}^4_f$ contain a compact set K that cannot be enclosed in any smoothly imbedded sphere $\mathbb{S}^3$, although this can be done with a continuously imbedded sphere. It follows from this that the spheres in $\mathbb{R}^4$ have a complicated (sawtooth) fractal structure near infinity.

Chapter 15

Soliton Particles

THE Yang–Mills equations are nonlinear, and therefore there is little hope of finding closed-form solutions. Such a statement seems quite plausible. Every student who has taken a course in differential equations will remember that the only differential equations for which a general solution is given in closed form are linear differential equations with constant coefficients. As often occurs in life, however, exceptions to the rule are sometimes are more interesting than the rules themselves. The wave equations for scattering have turned out to be an exception.

Let us digress a bit from quantum physics and talk about a phenomenon originally detected in hydrodynamics—the formation of a solitary wave. The concept of a solitary wave was introduced about 150 years ago by the British shipbuilding engineer John Scott Russell. In a paper presented to the British Society for the Advancement of Science, he gave an enthusiastic description of the phenomenon:

> I was watching the motion of a barge which a pair of horses was pulling at great speed along a narrow canal when suddenly the barge stopped sharply. But the mass of water it had set into motion in the canal was by no means stopped. Violently seething, it began to gather around the prow of the boat, and then suddenly, abandoning the boat, it rolled off ahead with tremendous speed, having taken the form of an isolated large mound—a roundish, smooth and sharply outlined mass of water which continued its path along the canal without any noticeable change of form or slackening of speed. I rode after it on horseback and when I caught it, it continued to roll forward at a speed of 8 to 10 miles an hour, preserving its initial form in the shape of a figure about 30 feet long and 1-1/2 feet high. The height of the water gradually decreased, and after pursuing it for one to two miles, I lost it in the windings of the canal. Thus, in the month of August 1834 occurred my meeting with such a peculiar and excellent phenomenon.

Russell's observation was by no means immediately accepted by the scientific community. Such authorities as G.B. Airy (1801–1892) and G.G. Stokes (1819–1903) doubted the possibility of the formation of a wave having constant shape and propagating above the level of the water. However, by the mid-1870's Joseph Boussinesq (1842–1929), in 1871, and Lord Rayleigh (John William Strutt, 1842–1919), in 1876, had confirmed Russell's results theoretically. In particular Boussinesq derived an equation that describes the propagation of waves in shallow water (that is, under the assumption that the amplitude of the wave is small in comparison with the depth of the water in the channel) taking account of dispersion and nonlinear effects. His equation describes the motion of waves moving in two directions (right and left) and has the form

$$u_{tt} - u_{xx} - (u^2/2)_{xx} + \frac{1}{u} u_{xxxx} = 0. \qquad (15.1)$$

Here and below we are using the standard notation:

$$u_t = \frac{\partial u}{\partial t}, \quad u_{tt} = \frac{\partial^2 u}{\partial t^2}, \quad u_x = \frac{\partial u}{\partial x}, \quad u_{xx} = \frac{\partial^2 u}{\partial x^2}, \quad \text{etc.}$$

The next most important event was the 1895 paper of D.J. Korteweg (1848–1941) and G. de Vries. The Korteweg–de Vries (KdV) equation also describes the propagation of waves in shallow water, but moving in only one direction. It is:

$$u_t + 6uu_x + u_{xxx} = 0. \qquad (15.2)$$

Although the KdV equation can be obtained by reduction from the Boussinesq equation (by retaining only the waves that move in one direction), there are many reasons why it has independent and more fundamental significance.

The solution of the KdV equation, which corresponds to a traveling wave, is not difficult to find by assuming that $u(x, t)$ has the form $u(x - \alpha t)$. If one changes to a coordinate system ξ moving at speed α, that is, $\xi = x - \alpha t$, the wave will appear to be stationary. In these coordinates, the KdV equation for a solitary wave will be an ordinary differential equation that is easy to solve.

One possible solution of solitary wave type when the asymptotic condition $\tilde{u}(\xi) \to 0$ as $\xi \to \pm\infty$ is chosen has the closed form

$$\tilde{u}(\xi) = 3\alpha \operatorname{sech}^2\left(\frac{1}{2}\sqrt{\alpha}\xi\right).$$

Solitary waves, which are localized lumps that preserve energy and move in space without changing shape, have received the name solitons. For a long time, solitary waves were treated as an unimportant piece of exotica encountered in two-dimensional problems of nonlinear waves. It was supposed that when two such waves collided they disintegrated completely; therefore there was no basis for considering soliton solutions to be sufficiently general.

The limited interest in the KdV equation is particularly noticeable when we trace the fate of the authors of this discovery. Diederik Johannes Korteweg was a prominent Dutch scholar, a professor at the University of Amsterdam, and a member of the Netherlands Academy of Science. Although his name is now known mostly for the KdV equation, his contemporaries regarded completely different works as his major achievements. As proof one need only note that this paper is not mentioned in his obituary in the *Proceedings of the Royal Society*, published by the Netherlands Academy of Sciences. Gustav de Vries, a student of Korteweg in whose dissertation (1894) this equation first appeared, was unable to obtain any work in a university at all. He spent his entire career as a Gymnasium teacher. The subsequent 60-year evolution of wave theory seemed to have justified such skepticism. But in 1955 something happened that forced a fundamental revision of this point of view. However, as often happens in life, there was at first no indication that such pivotal changes were on the way.

The beginning of this instructive story can be taken as a conversation of two leading specialists in the field of atomic and hydrogen weapons Enrico Fermi and Stan Ulam in the summer of 1952 at Los Alamos. Both of these famous scholars considered their military tasks completed and wished to return to more academic problems. As Ulam writes in his fascinating autobiography [U], Fermi and he wanted to find some substantive problem, in which they could apply the most powerful computer, MANIAC 1, created by their friend John von Neumann (1903–1957). With his characteristic ingenious intuition Fermi sensed the great value of nonlinear equations for future fundamental physical theories. As a result of extended discussions they chose a concrete and very difficult problem that had been posed at the beginning of the century by P. Debye—to explain the finite thermal conductivity of solid bodies.

Their model of a solid body was an anharmonic chain of n point masses connected by springs. The number of points in the first experiment was 64. Chains with two types of nonlinearity were considered:

$$\ddot{x}_i = (x_{i+1} + x_{i-1} - 2x_i) + \alpha(x_{i+1} - x_i)^2 - (x_i - x_{i-1})^2 \qquad (15.3)$$

and

$$\ddot{x}_i = (x_{i+1} + x_{i-1} - 2x_i) + \beta(x_{i+1} - x_i)^3 - (x_i - x_{i-1})^3, \qquad (15.4)$$

where x_i is the displacement of the ith point from its initial position, and α and β are coefficients of the quadratic and cubic force terms respectively, acting between adjacent masses. The coefficients α and β were chosen to be small. It was assumed that if the initial solution is chosen to correspond to the linear problem, in the shape of a simple sinusoidal wave (of low mode) or a linear combination of low modes, then under a nonlinear interaction the energy of the initial state would be uniformly distributed over all harmonics (modes of vibration) for large t (as $t \to \infty$).

However, the computations that they carried out together with the young physicist John Pasta led to very unexpected results. Instead of ergodic behavior of the system, they observed a quasi-periodic character: the energy did not thermalize. To the contrary, the energy contained in the lowest mode, after a certain redistribution among the low modes, again accumulated in the lowest mode (within a few percent), and then the process repeated.

The example of this remarkable work is the best possible illustration of the classical thesis that great discoveries in science always have small beginnings—an attempt to solve some interesting particular problem. The unexpected difficulties and paradoxes show in particular that they had succeeded in probing something fundamentally new. Such has been the case with nearly all the great discoveries: it suffices to recall the theory of relativity, quantum mechanics, and many others. So it was with the work of Fermi, Pasta, and Ulam. They had encountered the property of complete integrability of a nonlinear system. But this became clear only ten years later, after the work of FPU was published as a laboratory report from Los Alamos, and was connected with the names of two American physicists, Martin Kruskal and Norman Zabusky, specialists in plasma theory. They knew the work of FPU and were very much interested in the results obtained. It should be remarked that FPU chains become the equations of nonlinear strings in the continuous limit. For example, Eq. (15.4) becomes

$$u_{xt} + u_x u_{xx} + u_{xxxx} = 0.$$

If we set $u_x = v$, we obtain the KdV equation. It was therefore completely natural for Kruskal and Zabusky to begin with the KdV equation. As a result of the numerical experiments that they conducted some amazing properties of solitons in the KdV equation were discovered. Solitons were not destroyed by collisions; somehow they passed through one another, changing places. The picture of solitons colliding at different speeds is especially interesting: c_1 and c_2, $c_1 \gg c_2$. A soliton moving with great speed absorbs a soliton moving with less speed but then emits it again.

Two years later, in 1967, the Princeton physicists M. Kruskal, J. Green, C. Gardner, and R. Miura found a theoretical basis for the unusual properties of the KdV equation. They showed that KdV equations have an infinite series of conservation laws with a whole class of multisoliton solutions, that is, solutions of solitary-wave type $u(x - c_i t)$, moving with different speeds c_i. It was shown that the evolution of solitons in the KdV equations is described by a linear Schrödinger equation.

This remarkable paper can be regarded as the progenitor of the theory of solitons. A 1968 paper of P. Lax of the Courant Institute was of fundamental importance for the further development of the theory. Lax proposed a regular algebraic construction for finding integrable systems. His idea was so simple and basic that it can be described in a few lines.

Suppose given the evolution equation

$$u_t = K(u). \tag{15.5}$$

If there exists an operator representation $u \mapsto L_u$, where L is a symmetric operator such that the operators $L(t) = u(t)L_u u(t)^{-1}$ are unitarily equivalent, then the eigenvalues of the operator L_u are first integrals (conservation laws) of Eq. (15.5). It is easy to see that the unitary equivalence of the operators $L(t)$ is equivalent to the existence of a representation of the form

$$L_t = AL - LA = [A, L], \tag{15.6}$$

wbere A is a skew-symmetric operator, the generator of the one-parameter group L_t. For the proof it suffices to differentiate the one-parameter trajectory $L(t)$ with respect to t. Thus Lax' method consists of an operator-valued linearization of evolution equations. To do this it is necessary to find the operators L and A. For the KdV equation the corresponding linear operator is

$$L = \frac{d^2}{dx^2} + \frac{1}{6}u, \tag{15.7}$$

the operator A is given by

$$A = u\frac{d^2}{dx^2} - 3\Big(u\frac{d}{dx} + \frac{d}{dx}u\Big),$$

and the representation $L_t = [A, L]$ is precisely equivalent to the KdV equation.

In 1971, the Soviet scholars V. Zakharov and A. Shabat found yet another example of an integrable system—the nonlinear Schrödinger equation:

$$iu_t + |u|^2 u_x + u_{xx} = 0. \tag{15.8}$$

This equation is important not only from the point of view of applications. It describes various phenomena in nonlinear optics (self-focusing optical beams), plasma physics (the collapse of Langmuir waves), and other areas, but with the aim of extending the applicability of the Lax representation. In contrast to the KdV equation, in which the operators L and A are scalar-valued, in the case of the nonlinear Schrödinger equation L and A are 2×2-matrix-valued operators. Another result obtained in 1971 by Gardner, L. Faddeev, and Zakharov gave a precise meaning to the concept of integrability of nonlinear evolution equations. They showed that the KdV equation, in full analogy with classical mechanics, can be regarded as an infinite-dimensional integrable Hamiltonian system, that is, one can represent the trajectories of solutions of the KdV equation as quasi-periodic windings of an infinite-dimensional torus with Hamiltonian generated by one of the conservation laws (for the KdV equation). The coordinates of the "phase" torus are coordinates of "action-angle" type, and are constructed from the scattering data of the linear Schrödinger operator (15.7).

Subsequently analogous results were obtained for other integrable evolution equations. Thus a large class of nonlinear equations having the following properties was identified: the presence of N-solitonsolutions and an infinite number of conservation laws, the existence of a representation of Lax type, and a Hamiltonian structure. For this class of equations a regular procedure was found for finding solutions, which came to be known as the inverse scattering method, since the solutions were constructed from the scattering data of the corresponding linear operator.

In recent years integrable nonlinear equations seem to have poured from a cornucopia. The exceptionally intense work of mathematicians and physicists from many countries has brought about the formation of an articulated theory of integrable nonlinear systems that is rich in profound and interesting connections with the theory of Riemann surfaces, topology, and algebraic geometry. The most significant results are connected with the construction of multidimensional integrable systems and the identification of a class of solutions with periodic boundary conditions. We shall give just one example, in which the intertwining of different branches of the theory of integrable systems has led to the solution of a classical mathematical problem—Schottky's problem mentioned above (Chapter 4).

We begin by describing a class of two-dimensional integrable systems discovered in plasma physics back in 1970 by two Moscow physicists: B. Kadomtsev and V. Petviashvili:

$$\pm u_{yy} + (u_t + 6uu_x + u_{xxx})_x = 0. \tag{15.9}$$

These equations (the KP equations) describe a medium with weak dispersion (positive or negative depending on the coefficient of u_{yy}) and are just as universal in the two-dimensional case as the KdV equation is in the one-dimensional case. The KP equations have all the properties of a completely integrable system: the presence of N-soliton solutions, an infinite number of conservation laws, a generalized Lax pair, and the like.

In discussing integrable equations we have tacitly assumed up to now that the solutions are being sought in the class of rapidly decreasing functions (as $t \to \infty$). How does the statement of the problem change if we consider periodic boundary conditions?

In 1974 S. Novikov and P. Lax studied this problem in the case of the KdV equation and discovered the remarkable properties of solutions in the periodic case—their connection with Schrödinger operators with periodic potentials L_p. The spectrum of an L_p operator partitions the real line into a finite number of intervals called *zones*. Therefore the method of integrating the KdV equation in the periodic case came to be known as finite-zone integration. The solution of the periodic problem produced a number of new nontrivial properties of integrable systems. Even the definition of the analogue of a soliton turned out to be nontrivial. It turns out that the analogue of an N-solitonsolution is an N-zone potential.

By developing the method of finite-zone integration in the case of two spatial variables—the KP equation—Novikov's student I. Krichever found a very beautiful method of integrating equations of KP type, based on the ideas of algebraic geometry. The condition for integrability of two-dimensional systems can be represented as a generalized Lax representation:

$$L_t - A_y = [A, L]. \tag{15.10}$$

This turns out to be Krichever's key observation: the commutation condition (15.10) is equivalent to the existence of some "sufficiently complete" family of functions $\Phi(x, y, t, P)$ annihilated by these operators and defined on a Riemann surface with a distinguished point P. The functions Φ are defined by their analytic properties; in particular they have an essential singularity at P. The solutions of Eqs. (15.9) constructed using such functions can be expressed in terms of the Riemann θ-functions. In particular quasi-periodic solutions of the KP equation can be represented in the form

$$u(x, y, t) = q + 2\frac{d^2}{dx^2} \ln \theta(u_1 x + u_2 y + u_3 t + w), \tag{15.11}$$

where u_1, u_2, and u_3 are constant b-periodic vectors of meromorphic differentials (with a single pole at the point P), q is a constant, and w is a point of the Jacobian of the Riemann surface R_g of genus g.

We recall the basic concepts needed to understand formula (15.11), referring to any course in the theory of Riemann surfaces for details [V.3]. A basis of holomorphic differentials w_i is defined by a system of canonical cuts (cycles) a_i, b_i $(i = 1, 2, \ldots, g)$ on R_g with intersection matrix $a_i \cap a_j = b_i \cap b_j = 0$, $a_i \cap b_j = \delta_{ij}$. A canonical basis of forms can be chosen as $\int_{a_i} \omega_k = \delta_{ik}$ and $\int_{b_i} \omega_k = B_{ik}$, where the matrices B_{ik} are called the matrices of b-periods and satisfy the Riemann–Frobenius condition: B_{ik} are symmetric and have positive-definite imaginary part.

The integer lattice of vectors in complex space $\mathbb{C}^g$ with coordinates δ_{ik}, B_{ik} defines a complex torus $T(R_g) = \mathbb{C}^g/\{B, \delta_{ik}\}$, called the *Jacobian* of the surface R_g. The Riemann θ-function is constructed from the matrix B:

$$\theta(x_1, \ldots, x_n) = \sum_{m \in \mathbb{Z}^g} \exp\big(\pi i(Bm, m) + 2\pi i(m, x)\big),$$

where $(x, m) = \sum_{i=1}^{g} x_i m_i$.

In 1977 S. Novikov put forward a very interesting conjecture connected with the solution of Schottky's problem: *The class of θ-solutions of the KP equation* (15.9) *distinguishes the θ-functions of Riemann surfaces of genus g in the class of all θ-functions of the space $\mathbb{C}^g$.* The proof of Novikov's conjecture was obtained by T. Shiota in 1986.

The connection thus discovered between algebraic geometry and the theory of integrable systems has resulted in a profound mutual influence on both mathematical disciplines. In algebraic geometry it has led not only to the solution of classical problems or a new proof, but also to the construction of more efficient and geometrically intuitive constructions as a whole. One can even speak of a certain return to the traditional classical algebraic geometry of the late nineteenth and early twentieth centuries, which was strongly displaced by the abstract constructions of the postwar period (the 1950's and 1960's).

In the light of these recent achievements the undeservedly forgotten work of the 1920's gained a completely new appreciation. For example, the classification of the commutative algebras of scalar differential operators generated by the operators A and L, which plays a key role in Krichever's construction, was contained in the work of the British mathematicians J.L. Burchnall, T.W. Chaundy, and H.F. Baker. Results of equal interest, which have found their place in the theory of integrable systems, have been obtained by the French mathematicians René Garnier (1887–1984) and Jules Drach (1871–1941). Moving ahead somewhat, we remark that even the famous sine–Gordon equation was obtained in a geometric context back in the middle of the nineteenth century.

The motives for studying specific mathematical objects have been completely different in different periods, but all are united by one common property: *If the object being studied is substantial and "natural," it will not vanish without a trace, and once it has appeared, it will live many and quite unexpected lives.*

This thesis is fully applicable to the theory of solitons. New equations are constantly being discovered that admit exact solutions. The number of phenomena that are explained using solitons is also increasing. For example, it has been suggested that the large stable red spot in Jupiter's atmosphere is a soliton. Attempts to identify the blue spot in the atmosphere of Neptune with a soliton seem more doubtful, since it is more likely that this spot has already evaporated. The most interesting examples of solitons are vortices in liquid crystals and superfluid helium.

Specific problems associated with integrable models arise in field theory. The problem of finding relativistically invariant integrable systems, that is, invariant relative to transformations in the Lorentz group, is of obvious interest. For these systems one can consider the problem of quantization in a natural way. Unfortunately the majority of known examples are two-dimensional systems, with one space coordinate and one time coordinate. For field theory, where space-time is the Minkowski four-dimensional space, this is a serious constraint. Nonetheless, a number of properties of two-dimensional models, in particular the interaction of solitons are also of interest for realistic four-dimensional theories.

Several relativistically invariant two-dimensional integrable models have been found. The best-known such equation is called the "sine–Gordon equation,"

$$u_{tt} - u_{xx} = \sin u. \tag{15.12}$$

The name "sine–Gordon" now in general use for Eq. (15.12) was proposed by M. Kruskal around 1970 by association with the Klein–Gordon equation[1] (O. Klein and W. Gordon) but the equation itself is no less universal than the KdV equation in the number of different applications, and it has an even more ancient history.

Equation (15.12) appears in the work of geometers in the mid-1830's. For example the surfaces of revolution of constant negative curvature found by F. Minding (1838) and E. Beltrami (1872) are solutions of Eq. (15.12). Equation (15.12) seems to have been known even to Gauss.

Equation (15.12) arises as a very special case in the curious "applied" paper "On cutting fabric in the manufacture of clothing." In 1878 P.L. Chebyshev, in his paper "Sur la coupe des vêtements," read at the Association Française pour l'Avancement des Sciences, derived the equation for a grid on a surface having the following properties: In each quadrangle of the grid the lengths of the opposite sides are equal (such grids later came to be called *Chebyshev grids*). In the coordinates of a Chebyshev grid the element of length on the surface can be represented in the form

$$ds^2 = dx^2 + 2\cos\varphi\, dx\, dy + dy^2,$$

where $\varphi(x, y)$ is the angle between the grid lines and x and y are the arc lengths of the intersecting coordinate lines of the grid. Computing the Gaussian curvature of the surface, Chebyshev obtained the equation:

$$\varphi_{xy} = -K \sin\varphi. \tag{15.13}$$

[1]The replacement of Klein but not Gordon by "sine" seems to be for the sake of euphony and is not discriminatory.

In the case of a surface of constant negative curvature ($K = -1$) the grid lines become asymptotic lines and Eq. (15.13) becomes the sine–Gordon equation. Eq. (15.12) is one of the equations of the Gauss-Codazzi-Peterson system that define the embedding of a surface of constant negative curvature into $\mathbb{R}^3$. Later (1902) Hilbert, in analyzing Eq. (15.12), proved the impossibility of an isometric embedding of a complete surface of constant negative curvature (the hyperbolic plane) in $\mathbb{R}^3$.

A little earlier than Chebyshev the Swedish mathematician A. Bäcklund had begun the systematic study of the solutions of the sine–Gordon equation. He constructed a hierarchy of solutions of the sine–Gordon equation. Bäcklund's construction, which makes it possible to construct new solutions from known ones, later came to be known as the Bäcklund transform. This method is applicable to all integrable evolution equations and is widely used in modern research.

One of the first applications of the sine–Gordon equation in physics is connected with the problem of the motion of dislocations in a crystal (Ya. Frenkel, 1936). The sine–Gordon equation arose later in widely diverse areas of physics. For example, it describes the motion of a magnetic flux in Josephson superconductors and Bloch walls in magnetic crystals.

But the most important applications of the sine–Gordon equation are connected with elementary particle physics. The sine–Gordon equation can be interpreted as the simplest nonlinear model of field theory. The soliton solutions that exist in it describe particles. The sine–Gordon equation was proposed by the British physicist T. Skyrme (1922–1987) as early as 1958, as a nonlinear generalization of the Klein–Gordon equation. Skyrme is the author of a number of original ideas in field theory and nuclear physics that were insufficiently appreciated during his lifetime. He also has the honor (partly in collaboration with T. Perring, 1962) of having conducted the first numerical experiments on elastic collisions of solitons in the sine–Gordon model. Skyrme and Perring conducted essentially the same experiment that Kruskal and Zabusky had conducted on the KdV equation in 1965, and they obtained similar results, but did not develop their theory. As we now know, the sine–Gordon equation is also a completely integrable system. The study of the sine–Gordon equation from the point of view of the theory of integrable systems made it possible not only to include it in the general scheme, but also to construct a quantum analogue of it, which had great value for field theory. A unified approach was found in the search for solutions of quantum integrable systems: the quantum inverse method, which has a large sphere of applications and has revealed new and unexpected connections with such objects of mathematics as infinite-dimensional Lie algebras, knots, loops, and many others.

Practically all the achievements discussed in this chapter involve two-dimensional integrable systems. In field theory an exceptional number of integrable four-dimensional systems are known. The most interesting result is the proof that the duality equation is integrable. It was conjectured that the "pure" Yang–Mills equation itself is also an integrable system, but an analysis of several special cases has shown that this conjecture is wrong.

Solitons arise not only in integrable systems. For example, it is known that field theory with the Lagrangian

$$(\partial_t \phi)^2 + (\partial_x \phi)^2 - U(\phi),$$

where the potential $U(\phi)$, the familiar potential of Goldstone, $U(\phi) = (\mu^2/2)\phi^2 + (\lambda/4)\phi^4$, has a soliton solution; however, it does not appear to be a completely integrable system.

The 't Hooft–Polyakov monopole can also be interpreted as a soliton. Solitons appear in all theories with spontaneous symmetry breaking. Their properties are closely associated with the topology of the space of gauge fields. The application of the theory of solitons makes it possible to obtain important results on the structure of vacua and processes involving real particles. A whole series of interesting problems arise, but they are all too close to the frontier of science and have not matured enough for popular exposition.

Chapter 16

Knots, Links, and Physics

KNOTS are the oldest object of study in topology. They form the subject matter of the greater part of Listing's first treatise, *Vorstudien zur Topologie*. This venerable division of topology, which flourished in the late 1920's and early 1930's, withered and faded away after the war. Such things happen to entire subjects as well as to people. It was not that there were no outstanding unsolved problems left in knot theory. For example, the following natural problem remained open: *How can a knot be distinguished from its mirror image?* The main goal of researchers was to find systems of knot invariants that would provide a simple procedure to distinguish knots. The most important knot invariant, which made it possible to distinguish knots quite simply in a number of cases, was a polynomial invariant, more precisely the Laurent invariant, discovered by the American mathematician J.W. Alexander (1888–1971). The Alexander polynomial $\mathrm{Al}(t)$ is symmetric under the change of variable $t \mapsto \frac{1}{t}$ and does not make it possible to distinguish two knots that are mirror images of each other.

Despite some interesting particular results, knot theory lay on the periphery of modern mathematics until a 1984 event that shook the mathematical community. The New Zealander Vaughan Jones, while working in an entirely different part of mathematics—operator theory, or more precisely, the study of von Neumann algebras—discovered a new class of polynomial invariants that were much stronger than the Alexander polynomials. With the use of Jones' invariants it was possible to solve specific complicated problems, in particular, the problem of deciding for a large class of knots whether a knot and its mirror image are isotopic. In some special cases, for example, for the trefoil (Fig. 16.1), this had been known previously, but the proof had required special devices.

Jones' method of proof was as interesting as the result itself, combining such seemingly diverse areas of mathematics as knot theory and von Neumann algebras. The main intermediate link in Jones' construction was Emil Artin's theory of braids. I shall confine myself to a sketch of Jones' main ideas, referring the interested reader to his book [Jo].

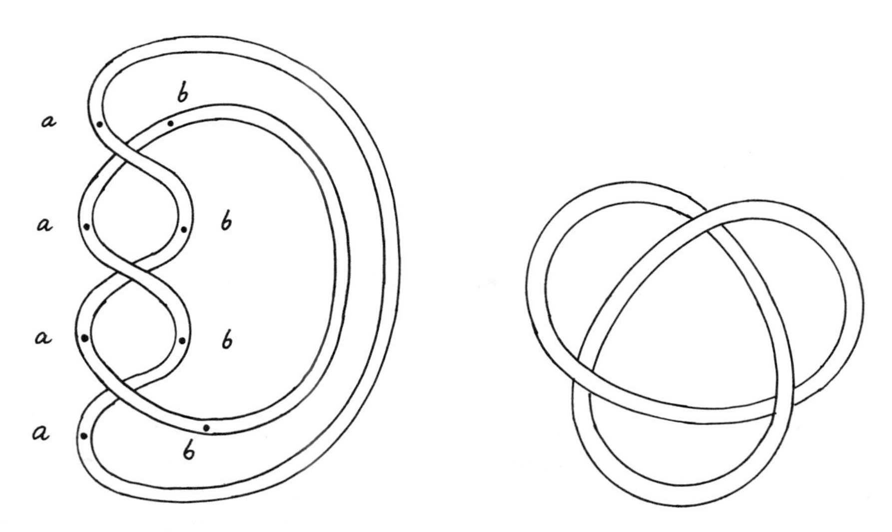

Figure 16.1: The trefoil knot.

16.1 Von Neumann Algebras

In the late 1920's, in connection with an analysis of the mathematical foundations of quantum mechanics, von Neumann began the study of algebras of operators with special commutativity properties. He studied an algebra $\mathcal{A}(\mathcal{H})$ of linear operators on a Hilbert space $\mathcal{H}$ satisfying the following conditions:

(a) *involutiveness* (for each operator A in the algebra $\mathcal{A}$, its adjoint A^* is also in $\mathcal{A}$);

(b) the algebra $\mathcal{A}$ is closed under the operation $*$ in the weak operator topology, that is, if A_n is a sequence of operators in $\mathcal{A}(\mathcal{H})$ and $\langle A_n x, y\rangle \rightarrow \langle Ax, y\rangle$ for some operator A and all vectors x and y in $\mathcal{H}$, then $A \in \mathcal{A}(\mathcal{H})$.

A von Neumann algebra $\mathcal{A}(\mathcal{H})$ is characterized by the interaction with its commutant $\mathcal{A}'(\mathcal{H})$, the algebra of operators that commute with $\mathcal{A}(\mathcal{H})$. It satisfies the condition

(c) $\mathcal{A}(\mathcal{H}) \cap \mathcal{A}'(\mathcal{H}) = \mathcal{C}(\mathcal{H})$, where $\mathcal{C}(\mathcal{H})$ is a commutative algebra.

The most important examples of von Neumann algebras are algebras whose center consists of scalars, that is, algebras $\mathcal{A}(\mathcal{H})$ for which $\mathcal{C}(\mathcal{H})$ consists of multiples of the identity operator (λE, where E is the identity of the algebra $\mathcal{A}(\mathcal{H})$). A ring $\mathcal{A}(\mathcal{H})$ with center $\{\lambda E\}$ is called a *factor*. The simplest example of a factor is the ring $\mathcal{B}(\mathcal{H})$ of all bounded operators on the Hilbert space $\mathcal{H}$. In the finite-dimensional case the ring of operators (matrices) on n-dimensional space $\mathbb{R}^n$ is the only example of a factor. This follows easily from the classical lemma of I. Schur. In infinite-dimensional spaces the situation is at once richer and more complex. An articulated and profound theory of factors was developed in a series of fundamental papers written by von Neumann partly in collaboration with F.J. Murray. A number of concepts and constructions introduced by von Neumann and Murray determined the development of the theory of factors for many years to come. The most important characteristic in the classification of a factor is its dimension. The dimension dim $(\mathcal{M})$ of a factor $\mathcal{M}$ has very exotic properties.

16.1.1 The Dimension of a Factor

Let us begin with a finite-dimensional example. In this case we have $\mathcal{M} \sim \mathcal{A}(\mathbb{R}^n)$, and the factor is numbered by a natural parameter, the dimension of the space $\mathbb{R}^n$ on which the full matrix algebra operates. In the infinite-dimensional case the analogous introduction of dimension encounters a natural obstacle, since all Hilbert spaces are isomorphic. The remarkable idea of von Neumann and Murray was to introduce analogues of finite-dimensional spaces in the infinite-dimensional situation. Such spaces are called *finite* spaces, and a concept of dimension can be defined for them by using a procedure known in high-school mathematics as the Euclidean algorithm or as division with remainder.

We need a few background concepts. Suppose given a factor $\mathcal{M}$ and a Hilbert space $\mathcal{H}$ on which it operates. $\mathcal{M}$ can be thought of as a subalgebra of $\mathcal{B}(\mathcal{H})$. Consider a sequence of Hilbert spaces $\mathcal{H}_i \subset \mathcal{H}$ *attached* to $\mathcal{M}$. This means that the mapping of the space $\mathcal{H}_i$ into $\mathcal{H}_j$ is carried out using projection operators P_i belonging to $\mathcal{M}$. The set of projections $\{P_i\}$ can be ordered. We shall say that the operator P_i *dominates* the operator P_j, denoting this fact $P_i > P_j$, if there exists an operator $u \in \mathcal{M}$ such that $uu^* = P_i$ and $u^*uP_j = u^*u$. The projections P_i and P_j are *equivalent* ($P_i \sim P_j$) if the following condition holds: there exists an operator u such that $uu^* = P_i$ and $u^*u = P_j$. It can be shown that the equivalence $P_i \sim P_j$ means that the conditions $P_i > P_j$ and $P_j > P_i$ both hold simultaneously.

For bounded operators the equivalence of the two projections P_1 and P_2 is equivalent to the condition that their ranges have the same dimension. But factor theory allows unbounded operators, requiring more delicate study. Murray and von Neumann proved that one of the two conditions $P_i > P_j$ and $P_j > P_i$ always holds for projections P_i belonging to a factor $\mathcal{M}$.

A projection P_1 is called *finite* if the conditions $P_2 < P_1$, $P_1 \sim P_2$ imply that $P_1 = P_2$; otherwise it is *infinite*. This terminology carries over naturally to the attached Hilbert spaces. A projection P is *minimal* if it dominates only the zero projection in $\mathcal{M}$. The existence of projections with the properties just listed makes it possible to separate factors into disjoint classes:

Type I: $\mathcal{M}$ contains a minimal projection P_{min};

Type II_1: $\mathcal{M}$ does not contain a minimal projection P_{min}, but it does contain a finite projection P_{fin};

Type II_∞: there is no P_{min}, but the factor $\mathcal{M}$ contains both finite and infinite projections;

Type III: The only finite projection in $\mathcal{M}$ is the zero projection.

Following Murray and von Neumann, one can define the dimension $d(\mathcal{M})$ of the factor $\mathcal{M}$, a numerical function $d : \mathcal{M} \to [0, \infty]$ that is uniquely determined by the following properties:

(1) $d(0) = 0$;

(2) $d\left(\sum_{i=1}^{\infty} P_i\right) = \sum_{i=1}^{\infty} d(P_i)$ if $P_i \perp P_j$ for $i \neq j$;

(3) $d(P_i) = d(P_j)$ if $P_i \sim P_j$.

It can be shown that the condition $d(P_1) = d(P_2)$ implies that $P_1 \sim P_2$, so that the different types of factors are uniquely determined by the function $d(P)$. If the function $d(P)$ is normalized, it assumes the following values:

I_n: $d = \{1, \ldots, n\}$;

I_∞: $d = \infty$;

II_1: $d \in [0, 1]$ (the whole interval);

II_∞: $d \in [0, \infty]$;

III: $d = 0$ or $d = \infty$.

The analogy with sets of numbers becomes especially obvious at this point. We shall demonstrate this using the example of type I_n. We denote the range of the function $d(P)$ by Δ. By definition of a factor of type I there is a smallest positive element α in the Δ. Let β be any element of Δ. For some integer n we have $n\alpha \leq \beta < (n+1)\alpha$. It easily follows from property (2) above that $\beta - n\alpha \in \Delta$. But $\beta - n\alpha < \alpha$. Since α is the minimal positive element, it follows that $\beta - n\alpha = 0$, that is, $\beta = n\alpha$. It also follows from property (2) that all the elements $\{\alpha, \ldots, (n-1)\alpha\}$ belong to Δ. Normalizing the quantity α, we obtain the spectrum of the dimension $d(P)$ for the factor I_n. Similar reasoning enables us to obtain the spectrum of $d(P)$ for factors of types II and III.

The most interesting factors for our purposes are those of type II_1. These factors arise most naturally in the theory of representations of discrete groups. For example, the algebra of operators generated by left translations (or right translations) on the group of rational automorphisms of the plane is a factor of type II_1. Here $G = \left\{\begin{pmatrix} a & b \\ 0 & 1 \end{pmatrix}\right\}$ is the group of triangular 2×2 matrices over the field of rational numbers. Although a description of the factors of type II_1 is incomparably more complicated than in the case of type I, these factors do possess one unifying property of decisive significance for their application in knot theory, namely the existence of a trace. The trace function $\mathrm{Tr}\,(A)$ is defined for all Hermitian operators $A \in \mathcal{M}$ and satisfies all the standard properties of a trace, including the properties $\mathrm{Tr}\,(\mathbf{1}) = 1$, $\mathrm{Tr}\,(AB) = \mathrm{Tr}\,(BA)$, and $\mathrm{Tr}\,(A^*A) > 0$ when $A \neq 0$.

Although a complete description of factors of type II_1 (up to isomorphism) has not yet been obtained, there is an important subclass of factors for which a complete description does exist. Those factors, which are closest to finite-dimensional factors, are called the *hyperfinite* factors. A hyperfinite factor $\widetilde{\mathcal{M}}$ is defined as the limit of an increasing sequence of finite-dimensional factors: $\widetilde{\mathcal{M}} = \{\mathrm{I}_n\}$. The limit is understood in the sense of the weak topology. Murray and von Neumann proved that among the factors of type II_1 there exists a hyperfinite factor that is unique (up to isomorphism). They also constructed examples of factors of type II_1 that are not hyperfinite. Subsequently continuous families of nonhyperfinite factors of type II_1 were found. To carry out a more detailed study of factors Murray and von Neumann introduced another numerical characteristic, the *coupling constant*, or, in modern terminology, the *index* of the factor $\mathrm{Ind}(\mathcal{M})$. The index measures the size of a Hilbert space $\mathcal{H}$ compared to the space $L^2(\mathcal{M})$ of square-integrable functions on $\mathcal{M}$. The index can be defined for any pair of factors $\mathcal{N} \subset \mathcal{M}$:

$$\mathrm{Ind}_{\mathcal{N}}\mathcal{M} = \dim_{\mathcal{N}}\big(L^2(\mathcal{M})\big).$$

The study of the properties of $\mathrm{Ind}_{\mathcal{N}}\mathcal{M}$ led Jones to an unexpected result.

Theorem 16.1 (Jones' Theorem).

1. *If* $\mathrm{Ind}_{\mathcal{N}}\mathcal{M} < 4$, *then there exists an integer* n, $n \geq 3$, *for which* $\mathrm{Ind}_{\mathcal{N}}\mathcal{M} = 4\cos^2 \pi/n$;
2. *For each* $n \geq 3$ *there exists a subfactor* $\mathcal{R}_0$ *of a hyperfinite factor* $\mathcal{R}$ *for which* $\mathrm{Ind}_{\mathcal{R}_0}\mathcal{R} = 4\cos^2 \pi/n$;
3. *For each real number* $r \geq 4$ *there exists a subfactor* $\mathcal{R}_0 \subset \mathcal{R}$ *with* $\mathrm{Ind}_{\mathcal{R}_0}\mathcal{R} = r$.

Jones' proof was based on an inductive construction of a sequential transition from the subfactor $\mathcal{N}$ to $\mathcal{M}$ using projection operators $P_{\mathcal{N}} : L^2(\mathcal{M}) \to L^2(\mathcal{N})$. Consider the sequence $\mathcal{M}_i$ defined as follows: $\mathcal{M}_0 = \mathcal{N}, \mathcal{M}_1 = \mathcal{M}, \ldots, \mathcal{M}_n = \langle \mathcal{M}, P_1, \ldots, P_{n-1}\rangle$, where P_i is the projection of $L^2(\mathcal{M}_i)$ onto $L^2(\mathcal{M}_{i-1})$. The projections P_i have the following properties:

$$\begin{aligned}
&(a)\quad P_i^2 = P_i = P_i^* \\
&(b)\quad P_iP_j = P_jP_i \ \text{ when } |i-j| \geq 2 \\
&(c)\quad P_iP_{i\pm1}P_i = \tau P_i, \ \text{ where } \tau = \mathrm{Ind}_{\mathcal{N}}^{-1}\mathcal{M} \\
&(d)\quad \mathrm{Tr}(wP_{n+1}) = \tau\mathrm{Tr}(w), \ \text{ where } w \text{ is} \\
&\text{the word generated by } \{\mathbf{1}, P_1, \ldots, P_n\}.
\end{aligned} \tag{16.1}$$

Jones proved this theorem in 1983. There remained one more step before it could be applied to knots. In taking that step Jones received help from Joan Birman, a prominent specialist in knot theory.

16.2 The Theory of Braids

Anyone who agrees that mathematical constructs have a relation to the real world must be amazed that the theory of braids did not appear until the twentieth century, especially considering that women's braids, horses' manes, ropes, and many other examples could have suggested a mathematical study of this subject. Serious doubts as to the correctness of such a pragmatic development of science forces me to confine myself to mentioning just these known facts. One way or another, the mathematical theory of braids arose in 1925 in Artin's paper "Theorie der Zöpfe." Subsequently, as often happens, it became clear that braid groups had occurred in earlier papers by F. Klein, R. Fricke, and A. Hurwitz.

16.2.1 The Group of Braids

Consider two sets, each consisting of n points. Arrange the first collection $(x_1, \ldots, x_n)$ on the plane $z = 1$ ($\mathbb{R}^2_1 \subset \mathbb{R}^3$) and the second $(x'_1, x'_2, \ldots, x'_n)$ on the plane $z = 0$ ($\mathbb{R}^2_0 \subset \mathbb{R}^3$). Join the n "upper" points to the n "lower" points by nonintersecting smooth curves ("threads"). Each such set of n threads (determined up to an isotopy) is called a *braid.* A multiplication can be defined on the set of braids, turning it into a group. The product of braids σ_1 and σ_2 is the braid σ_3 obtained by identifying the lower set of points of the braid σ_1 with the upper set of the braid σ_2. The identity element e and the inverse element σ^{-1} are determined with equal ease (see Fig. 16.2). Intuitively a braid can be thought of as a set of disjoint trajectories swept out by moving points $x_i(t)$. The inverse element is then obtained by simple time reversal. Denoting the group of braids with n threads by B_n, Artin proved that the group B_n is generated by $n - 1$ elements $\sigma_1, \ldots, \sigma_{n-1}$ (Fig. 16.2) satisfying the relations:

$$\begin{aligned} \sigma_i\sigma_j &= \sigma_j\sigma_i, \quad \text{when } |i-j| \geq 2, \\ \sigma_i\sigma_{i+1}\sigma_i &= \sigma_{i+1}\sigma_i\sigma_{i+1}. \end{aligned} \tag{16.2}$$

To close our chain of reasoning we need to explain the connection between braids, knots, and links. It is easy to see that each braid generates a certain knot or link. For this it suffices to "close off" the braid, that is, to connect the corresponding lower point with the upper point (Fig. 16.1). Alexander's result, proving that one can obtain any knot or link in this way, is much more complicated. Unfortunately this procedure is highly nonunique. The problem of determining the minimal number of threads needed to construct a given knot is extremely complicated and has not yet been solved in general.

The connection between knots and braids suggests studying which operations in the group of braids define isotopically equivalent knots. This problem was posed and solved by A.A. Markov, Jr., who showed that two types of transformations in the group of braids lead to equivalent knots. Let $\sigma \in B_n$. Then the following operations define equivalent knots:

$$\begin{aligned} &(1) \quad \sigma \mapsto \alpha\sigma\alpha^{-1}, \quad \text{where } \alpha \in B_n, \\ &(2) \quad \sigma \mapsto \sigma\sigma_n^{\pm 1}, \quad \text{where } \sigma_n \in B_{n+1}, \end{aligned} \tag{16.3}$$

and σ_n is the nth generator of B_{n+1}. We now construct a representation of the group B_n in the algebra generated by the projections $P_i : \{\mathbf{1}, P_1, \ldots, P_{n-1}\}$. Let us denote the element $tP_i - (\mathbf{1} - P_i)$ by g_i, where the parameter t is connected with τ by the

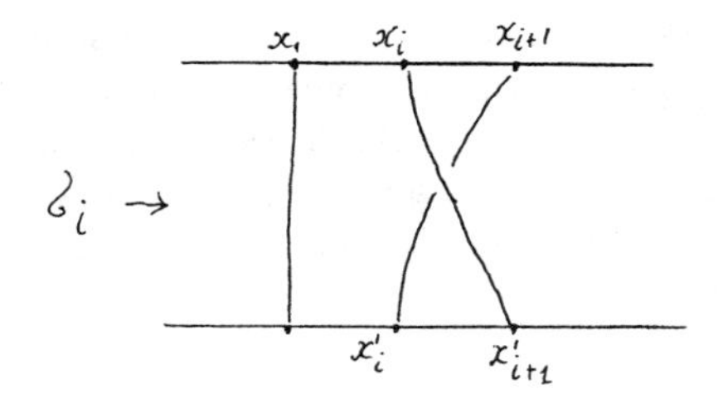

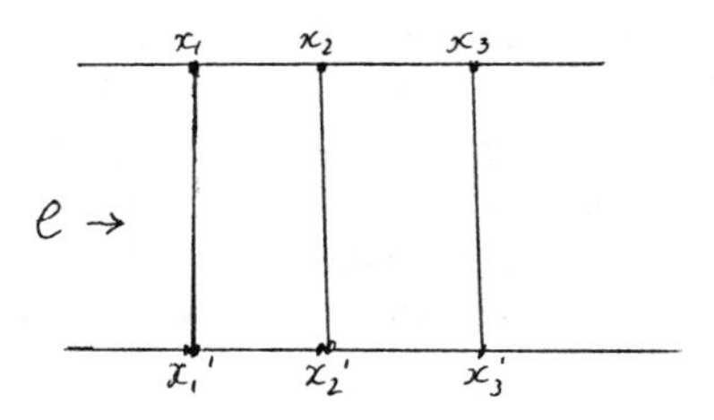

Figure 16.2: The braid group.

relation $\tau = 2 + t + t^{-1}$. Then relation (16.1) transforms into

$$\begin{aligned} &(1) \quad g_i^2 = (t-1)g_i + t \\ &(2) \quad g_i g_{i+1} g_i = g_{i+1} g_i g_{i+1} \\ &(3) \quad g_i g_j = g_j g_i \text{ when } |i-j| \geq 2 \\ &(4) \quad g_i g_{i+1} g_i + g_i g_{i+1} + g_{i+1} g_i + g_i + g_{i+1} + 1 = 0. \end{aligned} \tag{16.4}$$

The mapping $\varphi : \sigma_i \to g_i$ defines a representation of the group B_n for each value of t. If we take the trace of this representation $\mathrm{Tr}(\varphi(\sigma))$, it is easy to see that it is covariant relative to the transformations (16.4) and hence is an invariant not only of the braid σ but also of the knot or link $\hat{\sigma}$ corresponding to it.

Definition 16.1 The *Jones polynomial* of the link (or knot) L is the polynomial

$$V_L(t) = \left(-\frac{t+1}{\sqrt{t}}\right)^{n-1} t^{e/2}\mathrm{Tr}(\varphi(\sigma)),$$

where $\sigma \in B_n$, $L = \hat{\sigma}$, and the exponent e is the sum of the exponents of the word σ when it is decomposed into its generators σ_i.

For the trefoil, which is generated by the braid $\sigma_1^3 \in B_2$ (Fig. 16.1), the Jones polynomial is easy to compute by using the relations (16.4):

$$\mathrm{Tr}(\sigma_1^3) = \mathrm{Tr}((t^3+1)P - \mathbf{1}) = \frac{(t^3+1)t}{(t+1)^2} - 1,$$

and consequently

$$V_L(t) = \left(-\frac{t+1}{\sqrt{t}}\right)(\sqrt{t})^3\mathrm{Tr}(\sigma_1^3) = t - t^3 + t^4.$$

Thus $V_L(t) \neq V_L\left(\frac{1}{t}\right)$, and hence it follows immediately that the trefoil is not isotopic to its mirror image. The Jones polynomials distinguish a very large number of knots and their mirror images, but do not solve the problem completely, since there exist knots for which $V_L(t)$ and $V_L\left(\frac{1}{t}\right)$ are the same. The first example of this kind has nine intersections—the knot g_{42} from the book of Rolfsen [Ro]. Using Jones polynomials one can often obtain a lower bound for the number of threads in a braid generating the given knot. For computing the Jones polynomials it is convenient to use recursion (the skein relations). First introduced by Alexander in 1928, the skein relations were basically forgotten and rediscovered 40 years later by John Conway, who, by adding "initial conditions" to them, demonstrated their effectiveness in computing the Alexander polynomials of knots and links.

16.2.2 The Skein Relations

The transition from a link L_1 to a link L_2 can be regarded as a chain of local transformations of the following form:

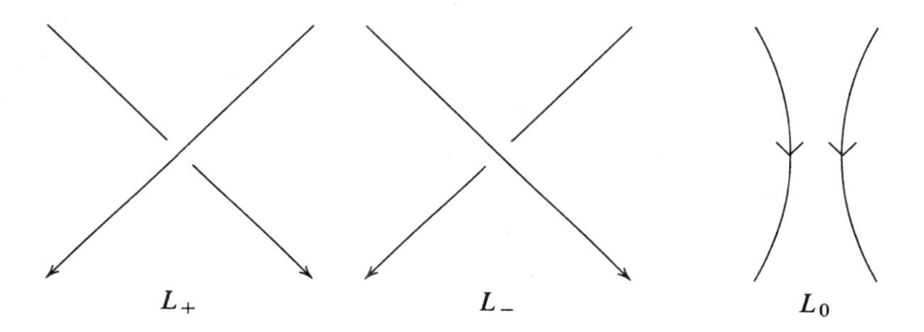

The Alexander polynomial Δ_L satisfies the relation

$$\Delta_{L_+} - \Delta_{L_-} = \left(\sqrt{t} - \frac{1}{\sqrt{t}}\right)\Delta_{L_0}.$$

If we adjoin Conway's "initial conditions"

(a) $\Delta_L(t) = 1$ if L is the trivial link

(b) $\Delta_L(t) = 0$ if L is a set of unlinked curves (circles), a *split link*,

The Alexander polynomial can be computed by induction for any link.

For Jones polynomials the skein relations have the form

$$\left(\frac{1}{t}\right)V_{L_+} - tV_{L_-} = \left(\sqrt{t} - \frac{1}{\sqrt{t}}\right)V_{L_0}.$$

Jones' result was immediately developed in several directions. Since the Jones polynomials do not subsume the Alexander polynomial, a two-parameter family of polynomials was constructed, containing the Alexander and Jones polynomials respectively in certain degenerate variables. Other polynomial invariants were also found that were not reducible to the Jones polynomial. A very interesting invariant is the Kauffman polynomial, which can be constructed immediately from the diagram of the link. The Kauffman polynomial is closely connected with lattice models of statistical physics. It is well adapted for studying the behavior of knots under a reflection and a change in the orientation of the ambient space.

But the most important events occurred in 1989, when two independent papers appeared. The first of these, written by E. Witten, was entitled, "Quantum field theory and the Jones polynomial." In that paper Witten pointed out the connection between the Jones polynomials and topological field theories. Let $\mathbf{M}^3$ be a three-dimensional manifold and L a link consisting of n circles $\{l_i\}$. On each manifold $\mathbf{M}^3$ one can define a field model with Lagrangian $\mathcal{L}$:

$$\mathcal{L} = k \int_{\mathbf{M}^3} \operatorname{Tr}\Big(A \wedge dA + \frac{2}{3} A \wedge A \wedge A\Big).$$

Here A is the Yang–Mills connection generated by the bundle over $\mathbf{M}^3$ with a certain gauge group G. The link L is connected with the functional

$$W_R(l_i) = \operatorname{Tr}_R P \exp \int_{l_i} A_i \, dx_i, \tag{16.5}$$

where R is an irreducible representation of the group G and P is the normal ordering needed to define the exponential of a connection with values in the Lie algebra of the group G. If the expression (16.5) is integrated with respect to the Feynman measure $\mathcal{D}A$, the integral (correlation function)

$$Z = \int_A \mathcal{D}A \exp(i\mathcal{L}) \Pi W_{R_i}(l_i)$$

defines the Jones invariants of the link L. Witten's idea made it possible to approach the computation of invariants of links by using the immense technical resources of quantum field theory. It also gave a completely natural solution to the problem of constructing the analogues of the Jones polynomials for knots and links imbedded in an arbitrary three-dimensional manifold. As is known, the properties of knots depend heavily on the topology of the ambient space, and Jones' original approach could not be easily carried over to manifolds different from the sphere $\mathbb{S}^3$.

Witten's ideas, possessing great heuristic power, were based on a number of physical structures with weak mathematical underpinnings. This was particularly true of the use of the Feynman measure. Nevertheless, nearly all of Witten's assertions were eventually proved. This turned out to be a highly nontrivial problem requiring the application of many abstract structures, including quantum groups and category theory.

Completely different ideas form the basis of the paper of the Moscow mathematician V. Vasil'ev. His point of departure was singularity theory, which studies the typical properties of smooth mappings. Vasil'ev invariants are constructed directly for the family of knots. Consider a set of smooth mappings $\mathbb{S}^1 \to \mathbb{S}^3$ having singularities or self-intersections. This set forms a hypersurface $\mathcal{D}$ in the space of mappings $\mathbb{S}^1 \to \mathbb{S}^3$

and is called the *discriminant.* The nonsingular points of the discriminant correspond to mappings with one point of transversal self-intersection, while the singularities are mappings whose derivatives have zeros, or mappings with nontransversal or multiple intersections. Any numerical invariant of the isotopic type of a knot can be defined using the discriminant. To be specific, to each nonsingular portion of the discriminant (that is, to any connected component of its set of nonsingular points), one can ascribe an index—the difference of the values of the invariant for the neighboring knots separated by this portion. This set of indices is not arbitrary, and to be well-defined it must satisfy a homological condition: the sum of the components taken with certain coefficients must be homologous to zero in the space K of all mappings $\mathbb{S}^3 \to \mathbb{S}^1$. Thus Vasil'ev numerical invariants are defined as locally constant functions on the space of imbeddings of $\mathbb{S}^1$ in $\mathbb{S}^3$, or, more precisely, as elements of the zero-dimensional cohomology group $H^0(K \setminus \mathcal{D})$. Along with the group $H^0(K \setminus \mathcal{D})$, the study of the higher cohomology groups $H^i(K \setminus \mathcal{D})$ is also of interest. The groups $H^i(K \setminus \mathcal{D})$ ($i \geq 0$) can be computed using a spectral sequence whose filtration is determined by the types and multiplicities of the singularities of the discriminant.

A fundamental problem of knot theory, the existence of a complete system of invariants in the Vasil'ev theory, reduces to determining the convergence of the spectral sequence. There is as yet no complete answer, but Vasil'ev's theory seems the most realistic route toward a solution. The results obtained in this direction confirm this point of view. In particular, it has been shown that both invariants of polynomial type (the polynomials of Jones, Alexander, and others) and the majority of classical invariants of the "pre-Jones" era (the Milnor coefficients and others) can be imbedded in the system of Vasil'ev invariants. Comparison with Witten's approach has also turned out to be promising. Perturbation theory, which was developed in topological field theory, makes it possible to obtain very explicit integral formulas for the Vasil'ev invariants of finite order. Thus, three formally distinct approaches to the theory of knots and links turned out to be closely connected with one another. The role of these methods in the study of the structure of three-dimensional manifolds seems still more important; in this area the famous Poincaré conjecture remains unresolved up to the present. Undoubtedly these theories, which connect such diverse structures as quantum groups, von Neumann algebras, Feynman integrals, and much else, conceal unexpected and profound discoveries.

Up to now we have discussed the purely mathematical aspects of knot theory, in which certain physical ideas have directly or indirectly turned out to have significant influence on the development of the theory. Let us now consider examples of the opposite process, some problems from traditional areas of physics in which applications of the theory of knots and links have real physical interest.

16.3 Condensed Matter Theory

How can knots and chains form in such orderly media as liquid crystals or superfluid liquids? Only a very indirect and incomplete answer can be given to this fundamental and highly nontrivial question. In the first place, experiments are known in which linked defects were observed; in the second place (and perhaps also in the first place) there is a fruitful operational principle in physics that requires the investigation of all possibilities not forbidden by the fundamental principles (the conservation laws, violation of causality, and the like). Among the systems in which linked defects have really been observed, we distinguish liquid crystals of nematic and cholesteric types. Other potential candidates are the quantum superfluid liquids ^{4}He and ^{3}He, in which linear defects of vortex thread and ring type have been observed. Unfortunately the study of vortices in ^{4}He and even more in ^{3}He is a difficult experimental problem, and vorticial rings have not yet been observed. But there are various indirect results, including computer models, showing that such configurations are possible. A number of observed effects are connected with the formation of linked vorticial tubes, for example, the occurrence of a chaotic (turbulent) mode in ^{4}He. As we have emphasized above (Chapter 12), liquid crystals, superfluid liquids, and neutron stars, which are so different in their physical characteristics, are very much alike from the mathematical point of view. All are determined by specifying the order parameter, and can be described by one or another of the Ginzburg–Landau equations. For that reason it is very natural to study the properties of linked defects in the context of some general scheme. From the topological point of view the theory of linked defects is the classical theory of links complicated by the introduction of the order parameter, which characterizes the corresponding thermodynamic phase. By developing these considerations, V. Retakh and the author proposed in 1984 a general method of describing linked defects. We shall illustrate the results of this paper using two examples having immediate physical applications.

16.3.1 Example 1

Consider the following problem. What is the structure of a system of topological invariants that would make it possible to decide whether one can decouple (using motions in $\mathbb{R}^3$) a system of linked closed curves (loops). A well-known invariant of this type is the classical Gauss linking coefficient k_G of two loops. But knowing this coefficient is not enough to solve the problem of decoupling. Well-known examples such as the Whitehead link and the Borromean rings (Fig. 16.3) show that the condition $k_G(l_1, l_2) = 0$ gives only a necessary condition for decoupling the two curves. For that reason, to solve the problem of decoupling, it is necessary to construct higher-order invariants. Such invariants are the high-order linking coefficients, which generalize the Gauss coefficients and were constructed in our paper. We shall now show what the high-order linking coefficients look like in the simplest case of a link of three curves $l = (l_1, l_2, l_3)$ embedded in $\mathbb{S}^3$.

We begin with the first-order linking coefficient $k_1(l_1, l_2)$ defined for any pair of linked curves l_1 and l_2; it is the algebraic sum of the number of intersections of the curve l_1 with a surface Z spanning the curve l_2 (with the orientation induced by l_2). This quantity is called the *intersection index* of Z and l_1, and is denoted $\operatorname{Ind}(Z, l_1)$. The number $k_1(l_1, l_2)$ is equal to the Gauss coefficient $k_G(l_1, l_2)$. For $k_G(l_1, l_2)$ there

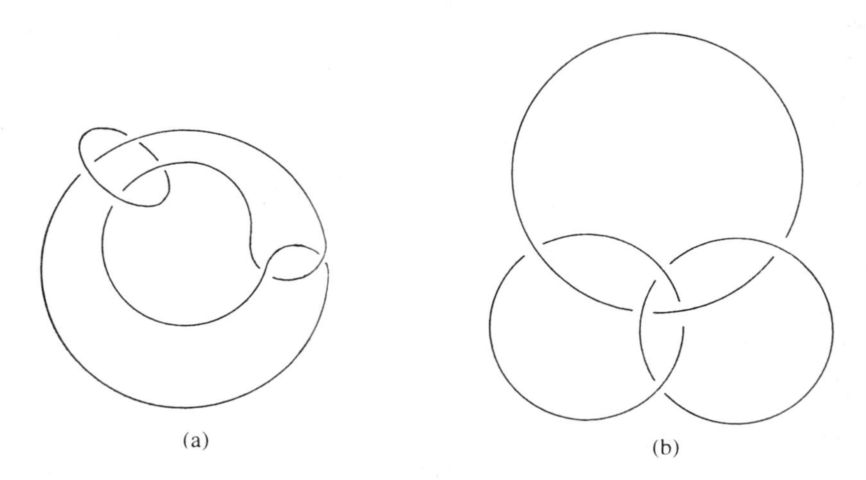

Figure 16.3: (a) The Whitehead link (b) Borromean rings.

exists an integral representation:

$$\int_{B_1} u_1 \wedge u_2 = -\int_{B_2} u_2 \wedge u_1 = k_G(l_1, l_2). \tag{16.6}$$

Here $B_i \subset \mathbb{S}^3$ are tubular neighborhoods of the curves l_i $(i = 1, 2)$, and u_i are the differential one-forms dual to the curves u_i in the sense of Alexander. The formulas (16.6) can be extended to the entire sphere $\mathbb{S}^3$ by introducing closed 2-forms v_i on $\mathbb{S}^3 \setminus l_i$ such that $\int_Z v_i = \mathrm{Ind}(Z, l_i)$, where Z is a (surface) cycle in $\mathbb{S}^3 \setminus l_i$. The 3-forms $u_1 \wedge v_2$ and $v_1 \wedge u_2$ are defined on $\mathbb{S}^3$, and

$$\int_{\mathbb{S}^3} u_1 \wedge v_2 = -\int_{\mathbb{S}^3} v_1 \wedge u_2 = k_G(l_1, l_2). \tag{16.7}$$

For a link consisting of n curves $l = (l_1, \ldots, l_n)$ we define $\bar{k}_1(l)$ as the quantity

$$\bar{k}_1(l) = \max_{1 \le i < j \le n} |k_1(l_i, l_j)|. \tag{16.8}$$

Consider the link $l = (l_1, l_2, l_3)$ with condition $\bar{k}_1(l) = 0$. Then there exist a one-form u_{12} on $\mathbb{S}^3 \setminus (l_1 \cup l_2)$ and 2-forms v_{12} and v'_{12} on $\mathbb{S}^3 \setminus (l_1 \cup l_2)$ such that $du_{12} = u_1 \wedge u_2$, $dv_{12} = -v_1 \wedge u_2$, and $dv'_{12} = u_1 \wedge v_2$. We now construct the differential forms

$$\begin{aligned} u_{123} &= u_{12} \wedge u_3 + u_1 \wedge u_{23} \\ v_{123} &= -v_{12} \wedge u_3 + v_1 \wedge u_{23} \\ v'_{123} &= u_{12} \wedge v_3 + u_1 \wedge v_{23}. \end{aligned}$$

The forms u_{123}, v_{123}, and v'_{123} are closed, and the forms v_{123}, v'_{123} are defined on $\mathbb{S}^3$. It can be proved that the integrals of the forms u_{123}, v_{123}, and v'_{123} over cycles are equal:

$$\int_{B_1} u_{123} = \int_{\mathbb{S}^3} v_{123} = \int_{\mathbb{S}^3} v'_{123} \tag{16.9}$$

and are integers, given a suitable normalization. These numbers are called the *linking coefficients* of $l = (l_1, l_2, l_3)$ of degree 2 and denoted $k_2(l)$. For an arbitrary link $l = (l_1, \ldots, l_n)$ the coefficient $\bar{k}_2(l)$ is defined in analogy with (16.8):

$$\bar{k}_2(l) = \max_{1 \leq i < j < k \leq n} |k_2(l_i, l_j, l_k)|.$$

Formulas of type (16.7) and (16.9) found immediate application in a number of problems of physics, in particular in magnetohydrodynamics. We shall consider one such problem later, but right now let us take up Example 2.

16.3.2 Example 2

Assume that we have a system of linked defects concentrated in a volume Ω. For definiteness we shall assume $\Omega \sim \mathbb{S}^3$. A specific physical system is determined by the order parameter Ψ and the domain of the order parameter (the space of internal states) V. In Chapter 12 we studied two types of liquid crystals in detail: the uniaxial nematic with $V \sim P^2$ and the cholesteric, for which the space V is isomorphic to the space $SO(3)/\mathbb{Z}_2 \times \mathbb{Z}_2$. The contractibility of a single defect is determined by the nontriviality of the fundamental group $\pi_1(V)$. But the study of a system of linked defects requires a study of the entire sequence

$$\pi_1(\mathbb{S}^3 \setminus l) \xrightarrow{\Psi} \pi_1(V). \tag{16.10}$$

The contractibility or noncontractibility of a link has great importance, since from the physical point of view a contractible link is not a defect. The condition of contractibility of links is determined by the linking coefficients and the group $\pi_1(V)$. The general theory is rather complicated, and so I shall confine myself to an illustrative elementary example. Suppose the link (defect) l consists of two curves l_1 and l_2. If $\Psi(l_1)$ is contractible in V, the coefficient $k_1(l_1, l_2)$ is even. This fact has a curious physical consequence. If we know from energy considerations that $k_1(l_1, l_2) \leq 1$, then $k(l_1, l_2) = 0$, and the curves are homotopically separable. Examples of such links have been observed in nematic liquid crystals in the experiments of Y. Bouligand.

16.3.3 Magnetohydrodynamics

Magnetohydrodynamics is the study of the motion of a conducting fluid or gas in a magnetic field. The equations of magnetohydrodynamics, which are a combination of the Maxwell equations and the equations of fluid mechanics are extremely complicated, and conceal not only physical but also topological profundities. We shall consider only one physically interesting problem, that of the reconnection of magnetic lines of force in a plasma. The basic equation that describes the variation of the magnetic field **H** taking account of the finite conductivity of the medium has the form

$$\frac{\partial \mathbf{H}}{\partial t} = \operatorname{curl}[\mathbf{v}, \mathbf{H}] - \operatorname{curl}(\nu \operatorname{curl} \mathbf{H}). \tag{16.11}$$

Here ν is the magnetic viscosity, and v is the velocity of the medium.

A natural question arises. Do there exist invariants that characterize the structure of the set of lines of force? In general the answer is simple: for $\nu = 0$ any topological invariant of the initial configuration of the lines of force is conserved. (This follows from a well-known theorem on frozen-in fields.) When $\nu \neq 0$ there are no topological invariants. However a realistic physical picture turns out to be much more interesting. If we consider the velocity of destruction of invariants at very small viscosities, we compute that at characteristic times of order $\ll \tau_d$ (where τ_d is the so-called *diffusion time*, $\tau_d = L^2\nu^{-1}$, L being the characteristic scale of variation of the configuration of the plasma and the field **H**) the topological structure is not destroyed completely. For that reason the question of conservation of topological invariants in this case is completely natural and correct.

This problem was solved by R. Taylor in 1974 under certain additional restrictions. He proved that the following quantity is conserved:

$$h = \int_{\Omega} (\mathbf{A} \cdot \mathbf{H})\, dv. \tag{16.12}$$

Here **A** is the vector potential of the field **H**, $\mathbf{H} = \operatorname{curl}\mathbf{A}$, and Ω is the total volume of the system on whose boundary $\mathbf{H}_\perp = 0$. Taylor's result can be restated as follows. A rapid restructuring of the topology of the lines of force for $t \ll \tau_d$ can change the linking of individual lines of force as a result of small-scale turbulence, but the total linking is conserved. Later B. Kadomtsev stated this result for the configurations of magnetic fields that vary in large-scale reconnections of lines of force. P. Sasorov and the author gave a rigorous definition of the reconnection process in terms of cutting and pasting of tubes of force. This definition makes it possible to give an exact meaning to the assertion that the single topological invariant h is conserved in the problem of rapid reconnection in simply connected domains. In multiconnected domains the magnetic flux is also conserved. Referring the reader interested in the physical problem to the paper [MS], I wish to spend a brief moment on the topological properties of invariants.

In the case of ideal magnetohydrodynamics Eq. (16.11) simplifies:

$$\frac{\partial \mathbf{H}}{\partial t} = \operatorname{curl}[\mathbf{v}, \mathbf{H}]. \tag{16.13}$$

In this case any invariant of the initial configuration of magnetic lines of force, which is conserved by virtue of the theorem on frozen-in fields, will be an integral of the motion. Hence it follows immediately that link invariants of type (16.7) and (16.9) will be conservation laws for Eq. (16.12). However, invariants of type (16.9), in contrast to h, are not conserved under reconnection.

It would be unfair to stop at this point without saying a few words in praise of the integral invariant h, which is often called, following Arnol'd, the *Hopf asymptotic invariant*. This name is far from a random choice. For an arbitrary field $\mathbf{A}$ the quantity h assumes real values, but if Ω is chosen as $\mathbb{S}^3$ and the field $\mathbf{A}$ is identified with the tangent field to $\mathbb{S}^3$, then the invariant h is the classical Hopf invariant of the bundle $f : \mathbb{S}^3 \to \mathbb{S}^2$, and h is integer-valued. The representation of h in the form (16.12) in this case was obtained by J.N.C. Whitehead as early as 1947. The Hopf invariant also makes possible another topological interpretation, proposed by Hopf himself. Suppose a regular mapping $f : \mathbb{S}^3 \to \mathbb{S}^3$ is given (the rank of f is 2), and x_0 and x_1 are regular points on $\mathbb{S}^2$. The preimages of the points x_0 and x_1 will be respectively curves l_0 and l_1 in $\mathbb{S}^3$. The *Hopf invariant* $h(f)$ of the mapping f is the linking coefficient of the curves l_0 and l_1. V. Arnol'd has shown that the Hopf asymptotic invariant can be interpreted as the average asymptotic linking number (over all pairs of lines of force). The latter is determined by the asymptotics of the linking of possibly nonclosed lines when they are extended indefinitely and closed in a nonsingular manner. This result can be generalized to higher linking coefficients.

The Hopf invariant arises in the most diverse problems of physics, from the Dirac monopole to the DNA ring structures, but that is a subject for another book.

In concluding this chapter and noticing with regret the large gaps and the sketchiness of the exposition, I comfort myself with the words of a classic of Russian literature, Koz'ma Prutkov: You cannot encompass the unencompassable. As compensation I refer the reader to a number of recently published books, surveys, and articles, where it is possible to obtain a very complete picture of the problems, both those touched upon here and those not touched upon [At], [Bi], [Ka], [Ko], [Mo], [MR], [N], [Tu].

Chapter 17

What Next?

IN the preceding chapters I have tried to give the reader an idea of some areas of physics in which topological methods have proved useful. I have tried to show that topological ideas are exceptionally close to modern (and not only modern) ideas in physics.

Although realistic applications of topology in physics began comparatively recently, several results can already be exhibited. If we consider topology in a broader mathematical context, including in it both algebraic and differential geometry (which is now completely justified), one can say that for modern physics topology is the same kind of defining structure that classical geometry was for Einstein's general relativity or group theory and Hilbert spaces were for quantum mechanics. Quantum field theory is still a promising area of applications for topology. Attempts to construct a unified field theory including all types of interaction lead to the study of high-dimensional bundles with a great variety of gauge symmetry groups.

At present the most promising construction is string theory and its generalization, membrane theory. In these theories particles are represented as elongated objects whose size is of the order of the Planck scale. Unfortunately, the possibility of experimental confirmation of the predictions of string theory are remote at present. For that reason physicists are studying strings and membranes as mathematical objects and discovering highly nontrivial properties in them. The results obtained in recent years in string theory have exerted a real effect on topology itself. Several of the latest discoveries—mirror symmetry and Seiberg–Witten invariants—have made great progress possible in the solution of classical problems of algebraic geometry and topology: an effective description of the moduli spaces of algebraic curves and the classifications of four-dimensional smooth manifolds. In both cases physical considerations connected with the occurrence of a certain symmetry-duality transform into precise mathematical assertions. In the first case this is the computation of the space of rational curves on special algebraic varieties (Calabi–Yau varieties); in the second case it is the discovery of invariants of spinor bundles on four-dimensional smooth varieties.

String theory provides a natural realization of ideas that connect topology and number theory. For example, when the coupling constants are small or large, the computation of the partition function of a string leads to nontrivial duality relations for the space of automorphic functions. The discrete symmetry that holds between electric and magnetic charges admits an extension to the group SL(2, $\mathbb{Z}$)) in the Yang–Mills theory,which in turn reflects the profound connection of string theory and the Yang–Mills equations. In studies of this whole circle of questions the most up-to-date mathematical methods harmonize beautifully with the classical structures, where Riemann's ideas occupy a position of high honor. In a number of other areas of physics the close connection with topology provides unexpected discoveries and new formulations of problems.

17.1 The Quantum Hall Effect

The quantum Hall effect is a remarkable example of a discovery showing that the most interesting things in science sometimes come about absolutely unexpectedly.

The classical predecessor of the quantum Hall effect was discovered by the British physicist E.H. Hall in 1879. What is curious is that it was published in a mathematical journal, the *American Journal of Mathematics*. The essence of the classical Hall effect is as follows: If a current is passed through a thin plate in a magnetic field perpendicular to the plate, a resistance (and conductivity) arises in a plane perpendicular to the moving current. The classical conductivity is Hall conductivity σ_{xy} (here σ_{xy} is the corresponding coordinate of the conductivity tensor), which is a function of the electron density and cannot be quantized.

The study of bulk semiconductors in strong magnetic fields over the past few decades has shown a monotonic variation in the resistance as the magnetic field is increased. Work in this field required sophisticated experimental technique, even though it seemed routine from the theoretical point of view. The announcement by the German physicist K. von Klitzing that he had discovered a discontinuous (quantum) character in the variation of Hall conductivity in two-dimensional silicon films, was all the more unexpected. His result was published in 1980 and awarded a Nobel prize in 1985, more than a century after the discovery of the classical Hall effect.

The motion of a stream of electrons in thin films ($\sim$ 50 Å) in a strong magnetic field and at sufficiently low temperatures ($\sim$ 1 K) can be considered two-dimensional. The energy spectrum of such a system is quantized—these are the so-called *Landau levels*. However, the condition of quantization of Hall conductivity does not at all follow from the condition of quantization connected with the behavior of an individual electron. Klitzig's experiment gave the quantum value of σ_{xy} with great precision (10^{-10}):

$$\sigma_{xy} = n\nu e^2/h.$$

Here e is the charge of an electron, h is Planck's constant, ν is a constant (the *filling factor*), and n is an integer. The importance of this discovery is difficult to overestimate when one takes into account that the quantity e^2/h is the fine structure constant—a fundamental constant in quantum electrodynamics.

The occurrence of quantum levels in Hall conductivity is itself of great interest from the point of view of fundamental physics. These quantum levels are not connected with the geometry of the image nor with the occurrence of impurities, and undoubtedly are of topological nature. Theoreticians had not yet managed to think up any explanation for the integer quantum Hall effect in 1982, when a new surprise awaited them. A group of experimenters at Bell Laboratory (Cherry Hill, New Jersey), consisting of D.C. Tsui, H.L. Störmer, and A.C. Gossard, discovered a fractional quantum effect in much thinner films, that is, the occurrence of levels of the form:

$$
\begin{aligned}
n &= 1/3,\ 2/3,\ 4/3,\ 5/3 \\
n &= 1/5,\ 2/5,\ 3/5,\ 4/5,\ 7/5,\ 8/5 \\
n &= 2/7,\ 3/7,\ 4/7,\ 5/7,\ 10/7,\ 11/7.
\end{aligned}
$$

This phenomenon came to be known as the *fractional quantum Hall effect.*[1]

Here again, although the independence of these quantum numbers from geometry and other similar effects seems to indicate that they are of a topological nature, the occurrence of fractional charges has not yet found a satisfactory explanation. Moreover there exist completely different explanations of the occurrence of integer and fractional charges. Even the occurrence of fractions with different parities in the denominator has no unified description.

While the theory of the integer quantum Hall effect finds a sufficiently natural explanation in the context of filling of the Landau levels, and the topological invariants arise as a result of integrating the wave functions of the system over basis cycles determined by boundary conditions and gauge transformations, the fractional hall effect does not fit into this system. Strenuous efforts are being undertaken in an attempt to explain the quantum Hall effect on the basis of modern topological field theories, the so-called Chern–Simons models. From the mathematical point of view these theories have a natural connection with such fundamental mathematical structures as braid groups, knots, and the like, in which fractional invariants have a natural explanation. But final success on that road remains distant.

It is interesting that the development of the experimental technique is advancing in a direction that makes it possible to study not only two-dimensional electron "fluids" and "gases," but also "quasi-one-dimensional" (localized in one dimension) and even zero-dimensional ones (quantum dots). The study of such systems has begun only recently and is furnishing an equal number of surprises.

17.2 Quasicrystals

The brief time interval from 1980 to 1985 was marked by several fundamental discoveries. The next in importance after the quantum Hall effect just mentioned and high-temperature superconductivity was the discovery of quasicrystals. In a paper

[1]Charges with even denominators were found later ($n = 3/2,\ 5/2$). For this discovery, H.L. Störmer, D.C. Tsui, and theoretician R. Laughlin, were awarded the 1998 Nobel Prize in physics.

published in 1984 in the *Physical Review Letters*, D. Schechtman, I. Blech, D. Gratias, and J. Cahn announced that they had observed icosahedral symmetry in certain metallic alloys (in particular Al–Mn and Al–Mn–Si). A short while later the observation of decagonal symmetry was announced. (Systems with 8th- and 12th-order axes of symmetry were subsequently found.) As is well-known in crystallography, crystals with 5th-order, 8th-order, 10th-order and 12th-order, etc., axes of symmetry do not exist, since such symmetries are inconsistent with the translation-invariance of the crystal lattice. Thus physicists had discovered a new structure having high local symmetry, but lacking spatial periodicity. Structures of this type were called *quasicrystals*.

The discovery of quasicrystals could not have occurred at a more opportune time, if one can use such an expression to characterize an outstanding discovery. Theoreticians had already constructed models of solid bodies with quasiperiodic tilings of the plane. In particular the famous Penrose tilings with two types of rhombi having angles of $144°$–$36°$ and $108°$–$72°$ can be regarded as a quasicrystal structure, while in solid state physics the so-called *incommensurable* structures were being studied. However, this research was lacking in motivation, which appeared in connection with the discovery of a real physical object realizing the theoretical mathematical structures.

As a result of extraordinarily intensive and fruitful research in which physicists, crystallographers, and mathematicians all participated, it became possible to explain the deep properties of quasicrystals having general scientific interest. If we confine ourselves to only theoretical questions, the original problem was to classify quasicrystals. It turned out that the majority of quasicrystals (but not all) can be obtained by projecting a multidimensional "crystal" onto three-dimensional space. The idea of the construction is as follows. Consider the six-dimensional Euclidean space $\mathbb{R}^6$, in which a representation of the icosahedral group I is acting. It is known that the representation I can be decomposed into two invariant three-dimensional representations acting in $\mathbb{R}^3$ and an orthogonal copy of it $\mathbb{R}^{3\perp}$. We embed the space $\mathbb{R}^3$ in $\mathbb{R}^6$ at an "irrational" angle. This means that the intersection of $\mathbb{R}^3$ with the integer lattice $\mathbb{Z}^6$ in $\mathbb{R}^6$ consists of the origin alone. The projection of $\mathbb{Z}^6$ onto $\mathbb{R}^3$ defines a tiling of $\mathbb{R}^3$, and that tiling forms a three-dimensional quasicrystal. In the case of a two-dimensional quasicrystal the ambient space is $\mathbb{R}^5$, in which a representation of the group $\mathbb{Z}_5$ acts. The group $\mathbb{Z}_5$ acts by cyclic permutations of the basis vectors $\mathbf{e}_1, \ldots, \mathbf{e}_5$ in $\mathbb{R}^5$. The representation of the group $\mathbb{Z}_5$ can be decomposed into the direct sum of three irreducible representations generated by the characters of the group $\mathbb{Z}_5$: $\{\exp(\pm 2\pi/5), \exp(\pm 4\pi/5), 1\}$. The two-dimensional plane spanned by the two vectors of the first representation is irrationally embedded in $\mathbb{R}^5$, and the projection of the integer lattice $\mathbb{Z}^5$ onto it defines a quasicrystal tiling. In this way one can obtain a Penrose tiling (Fig. 17.1). This construction was invented by the Dutch mathematician N.G. de Bruijn several years before the discovery of quasicrystals. Various approaches have been proposed to describe all possible quasicrystal tilings of the plane and three-dimensional space, including both geometric constructions based on certain rules for pasting elementary cells together (local rules) and more algebraic methods connected with the generalization of the concept of a crystallographic group.

Figure 17.1: A Penrose tiling.

Yet another area of research is connected with the description of the structure of defects in quasicrystals. This problem turns out to be significantly more complicated than in ordinary or liquid crystals. The methods of homotopy theory are not directly applicable in the theory of quasicrystals. They can give only a very rough estimate of the possible types of defects. The main difficulties involve the fact that there is no natural way to distinguish the internal degrees of freedom from the spatial degrees of freedom in a quasicrystal; in mathematical language this means that the order parameter space is locally dependent on the point of physical space. Attempts to get around this difficulty are leading to interesting geometric structures. In particular M. Kléman is the author of the interesting idea of describing defects in a quasicrystal by passing to hyperbolic geometry. Nevertheless, the problem of describing defects in a quasicrystal cannot yet be considered solved.

At present there seems to be a lull in the physics of quasicrystals. Experimental data are being compiled, rather large specimens of quasicrystals have been grown, and their various physical characteristics (electric conductivity, magnetic properties, and the like) are being studied. However, there is no doubt that the connection between quasicrystals and profound mathematical and physical structures will provide this area of research with a bright future.

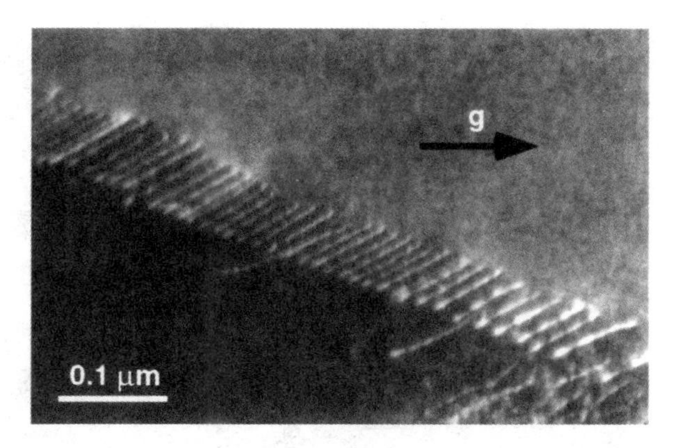

Figure 17.2: Disclination in quasicrystals.

17.3 Membranes

In recent years there has been great activity in so classical an area of mathematics as the theory of minimal surfaces. Connections have been discovered between that subject and the theory of integrable systems, some old problems have been solved, and new examples of minimal surfaces have been found. Computer graphics has made it possible to obtain a very visualizable representation of minimal surfaces. Progress in this area has been motivated to no small degree by the discovery of new physical phenomena. The most interesting and still little-studied systems are lyotropic liquid crystals, in which lamellor structures have been discovered of complicated shape with various types of symmetry (cubic, smectic, and the like). In particular vesicles have been constructed made of organic molecules, more precisely phospholipid bilayers, which from the mathematical point of view can be regarded as a two-dimensional surface. The shape of the surface is determined by the bending energy, which depends on the mean curvature H of the surface:

$$\mathcal{E} = k \iint_M H^2 \, dS, \quad H = \frac{r_1 + r_2}{2}, \tag{17.1}$$

Here r_1 and r_2 are the principal curvatures of the surface M, and k is the bending rigidity.

The problem of determining the shape of the surface reduces to finding the minimum of the functional (17.1). Here both compact and noncompact surfaces are of interest. In the compact case it is natural to study the extremal problem in the class of surfaces of fixed genus, by adding to (17.1) a term proportional to the Gaussian curvature

$$k_G \iint K \, dS. \tag{17.2}$$

A functional of type (17.1) is known in modern mathematics as a *Willmore* functional, although problems of this type had been studied as early as the 1920's. This functional is connected with important variables of the problem. For example, the Willmore conjecture that for toroidal surfaces the functional (17.1) has a global minimum on the flat torus (the Clifford torus): $\mathcal{E}_{\min T} = 2\pi^2$.

The majority of Willmore surfaces (but not all) can be obtained by stereographic projection of minimal surfaces embedded in the three-dimensional sphere. For physics it is important to determine the number of independent parameters in the space of Willmore surfaces of fixed genus—the number of independent conformal modes—since they are connected with the modes of thermal fluctuations of the surface. It can be shown that conformal modes are determined by the Riemann moduli of the surfaces. A detailed study of this question is of great interest.

Membranes of noncompact type also occur in nature. They form a system of periodic minimal surfaces (planes, for example) joined by tubes. It is remarkable that the first examples of such surfaces were constructed by Riemann himself in the early 1860's.

Besides the traditional problems of the theory of minimal surfaces, curious nonstandard mathematical problems arise in membrane theory. For example, *describe the behavior of the membranes for which the Gaussian curvature is not fixed, but only some distribution of it is given.* This is essentially a matter of a two-dimensional realization of the ideas of Clifford and Hawking in the theory of gravitation. This two-dimensional "foam" forms a special phase, the *sponge phase*. The study of membranes is important not only for physics, but also for biology. Biological membranes play a decisive role in the life of a cell. There is reason to hope that research in membrane theory will stimulate new mathematical problems as well.

Unexpected topological applications are being discovered in seemingly well-studied areas. In the study of Fermi surfaces of normal metals in magnetic fields **H**, S. Novikov and his students have discovered nontrivial topological properties of dynamic systems on surfaces. These results have immediate physical consequences. In particular it has been proved that in the situation of general position, when the quasiclassical trajectories of electrons are not closed, there exists a direction q, orthogonal to the field **H**, along which conductivity tends to zero. It is vital that all this is part of a more general mathematical theory—the theory of multivalued functionals. The theory of multivalued functionals is a generalization of Morse theory and has broad applications.

We have taken our last example from the work of the French physicists A. Joets and R. Ribotta, who are studying the optical properties of convective flows in liquid crystals. In polarized light one can observe a complicated behavior of defects in the transition from the laminar mode to the turbulent mode. Besides the study of the dynamics of this process, in which dissociation and reconnection of defects plays an essential role, the study of the caustics formed by reflected light has independent interest. From the mathematical point of view this leads to the study of caustics with a symmetry group. Here new problems arise, in which the methods of singularity theory and dynamic systems may be effectively applied.

There are many other interesting topics in which topology and physics complement each other beautifully. One may mention the latest research in the statistics of knots and the applications of knot theory to the analysis of the structure of DNA, papers on quantum gravity, and a number of others; but it seems to me that the examples just discussed will suffice to convince the reader that the romance of topology and physics is fated to have a long and happy life.

Chapter 18

A Brief Historical Survey

THE reader may be interested in the history of the origin of the series of ideas which have been discussed in this book.

A large part of the topological concepts used in examples considered here are far from new. Classification of two-dimensional surfaces was completed in the early 1920s. The hedgehog theorem (a result of the index theorem for critical points of vector fields) was proved by Poincaré for two-dimensional surfaces, and the multidimensional case was cleared up in 1926 by the Swiss topologist H. Hopf (1894–1971). The concept of homotopy groups and fiber bundles appeared in the mid 1930s.

At the time physics was proceeding in fundamentally different directions. Physicists were concerned with quantum mechanics of nuclei, scattering processes, and related matters. The analytic apparatus of theoreticians included the standard divisions of mathematics: the theory of functions of one complex variable, the apparatus of special functions, operations with matrices, and so forth. It was only the energetic activity of such enthusiasts as E. Wigner (1902–1995) and H. Weyl (1885–1955) that paved the way to apply group-theoretic methods.

After the war the situation began to change somewhat. New tendencies in theoretical physics came noticeably closer to the new concepts in topology; however, this convergence passed completely unnoticed by both sides. An illustration of this oversight is the introduction of connections into fiber bundles. In general form this concept appeared in the works of the French mathematician Ch. Ehresmann (1905–1979) in 1950 and in physics in the form of Yang–Mills gauge fields in 1954. As one of the authors, the Nobel laureate C. Yang, recounted, “At the time we were interested in equations and did not think about their geometric interpretation.”

Yang became acquainted with the mathematical theory of fiber bundles only 20 years after the appearance of his work with Mills. Speaking in 1979 at a symposium dedicated to one of the greatest contemporary geometers, S. Chern, Yang shared his reminiscences:

> Around 1968 I realized that gauge fields, both non-Abelian and Abelian, can be formulated in terms of nonintegrable phase factors, that is, path-dependent group elements. I asked my colleague Jim Simons about the

> mathematical meaning of these nonintegrable phase factors, and he told me they are related to connections with fiber bundles. But I did not then appreciate that the fiber bundle was a deep mathematical concept. In 1975 I invited Jim Simons to give the theoretical physicists at Stony Brook a series of lectures on differential forms and fiber bundles. I am grateful that he accepted the invitation and I was among the beneficiaries. Through these lectures T.T. Wu and I finally understood the concept of nontrivial bundles and the Chern–Weil theorem, and realized how beautiful and general the theorem is. We were thrilled to appreciate that the nontrivial bundle was exactly the concept with which to remove, in monopole theory, the string difficulty, which had been bothersome for over forty years [that is, single threads emanating from a Dirac monopole]... . When I met Chern, I told him that I finally understood the beauty of the theory of fiber bundles and the elegant Chern–Weil theorem. That non-Abelian gauge fields are conceptually identical to ideas in the beautiful theory of fiber bundles, developed by mathematicians *without reference to the physical world*, was a great marvel to me. In 1975, I mentioned this to Chern. I said, "This is both thrilling and puzzling, since you mathematicians dreamed up these concepts out of nowhere." Chern immediately protested, "No, no, these concepts were not dreamed up. They were natural and real."[1]

The development of physics and mathematics is proceeding along independent paths. Every science has its own internal motivating forces. Nonetheless, it turns out that for each fundamental theory in physics there is a corresponding specific mathematical structure.[2]

Physics	*Mathematics*
Special relativity	Four-dimensional space-time
General relativity	Riemannian geometry
Quantum mechanics	Hilbert space
Electromagnetism and non-Abelian gauge fields	Fiber bundles
String theory	Moduli spaces on Riemann surfaces

Complex interactions between mathematics and physics have also been observed in more modest situations. In his classical 1907 paper on elasticity theory, "On the equilibrium of elastic multiconnected bodies," the famous Italian mathematician V. Volterra (1860–1940) showed that the formation of internal stresses (singularities) in a solid body depends on the order of connectivity of the medium. Volterra clearly recognized the topological origin of singularities in a solid body. He obtained a series

[1] C.N. Yang, "Chern Symposium," June 1979 (preprint CERN TH 2725 [1979]); "Magnetic Monopoles, Gauge Fields, and Fiber Bundles," (preprint ITP/SB 77–14).

[2] This table of comparisons is borrowed from the paper of C.N. Yang presented to the conference dedicated to the sixtieth birthday of R. Marshak. Only the last line was added by the present author.

of purely mathematical results which are interesting for the topology of surfaces. The work of Volterra enjoyed a wide reputation among specialists; however, its topological basis did not find subsequent development in the physics of a solid body.

When one reads a work on liquid crystals (for example, the outstanding book of P. de Gennes, *Liquid Crystals*), topological methods for describing many phenomena suggest themselves. The first accurate applications of homotopy theory appeared only in 1976. In field theory, this occurred two years earlier, although the paper of D. Finkelstein and C. Misner "Some new conservation laws," was published as early as 1959, in the *Annals of Physics*. This paper applied topological methods to study the structure of space-time, and topological charges with the help of homotopy groups were introduced. One can only regret that this paper did not attract more attention.

The well-known physicist and mathematician F. Dyson devoted an interesting article to the interactions of physics and mathematics. Its title, "Missed Opportunities," gives an idea of Dyson's views on the situation.[3] The work of the last decade shows that mathematicians and physicists have learned the lessons of history, and a closer relationship already has given promising results. This book has discussed only a small part of such accomplishments, but the examples presented are sufficient to appreciate the wise words of the famous French mathematician Jacques Hadamard, "He who would unlock secrets should not lock himself away in one area of science but should maintain connections with its other areas as well."[4]

[3]F.J. Dyson, *Bull. Amer. Math. Soc.*, **78** (1972), p. 635.

[4]Hadamard, J. *Essai sur la psychologie de l'invention dans le domaine mathématique*, Blanchard: Paris, 1959.

Bibliography to Part II

This list of sources should help the interested reader learn about the themes touched on in Part II of this book in greater detail. The list includes monographs, popular books, reviews, and some original papers whose results are not yet reflected in the popular literature. There is also a historical section where classic works are noted.

I. Classic Works

1. Poincaré, H., "Analysis situs," *Journal de l'Ecole Polytechnique* **1** (1895), 1–121. [This work, together with its five appendices, defined the development of topology for many years. All articles are reprinted in the collected works (*Oeuvres*) of Henri Poincaré.]

2. Poincaré, H., *Oeuvres*, vols. I–XI, Gauthier–Villars, Paris, 1916–1956.

3. Volterra, V., "Sur l'équilibre des corps élastiques multiplement connexes," *Ann. Ecole Norm.*, **24** (1907), 401–517; Volterra, V., "The application of topological ideas in the theory of elasticity," *Opere Mathematiche*, 3 vols. Rome, 1956.

II. Topology

1. Milnor, J.W., *Topology from the Differentiable Viewpoint*, The University Press of Virginia, Charlottesville, 1965.

2. Wallace, A.H., *Differential Topology*. New York: Benjamin, 1968. [Two small books which are excellent complements to one another. Designed for mathematics students in lower courses. Presented without excessive formalism are the fundamentals of the topology of manifolds, including the Poincaré–Hopf theorem on indices of vector fields. Much attention is devoted to the theory of critical values of functions, Morse theory.]

3. Dubrovin, B.A., Novikov, S.P., and Fomenko, A.T., *Modern Geometry—Methods and Applications*, Pt. I, 1984; Pt. II, 1985; Pt. III, 1990, Springer-Verlag, New York. [This is a fundamental course, including basic facts from homology and homotopy theory, oriented toward the interests of a theoretical physicist.]

4. [Mo] Monastyrsky, M., *Topology of Gauge Fields and Condensed Matter*, Plenum, New York, 1993. [A course of topology aimed at physicists. It includes applications of topology in the theory of liquid crystals, superfluid liquids, and Yang–Mills fields.]

5. [Th] Thurston, W.P., *Three-dimensional Geometry and Topology*, Vol. 1, Princeton University Press, Princeton, 1997.

6. [DK] Donaldson, S., Kronheimer, P., *The Geometry of Four-manifolds*, Oxford University Press, Oxford, 1990.

7. [FU] Freed, D., Uhlenbeck, K., *Instantons and Four-manifolds*, 2nd ed, Springer-Verlag, New York, 1988.

8. [D] Donaldson, S.K., "The Seiberg–Witten equations and 4-manifold topology," *Bull. Am. Math. Soc.*, **33**, No. 1 (1996), 45–70. [This article discusses the Seiberg–Witten invariants and their connection with spinor bundles on 4-dimensional manifolds.]

9. [GY] Greene, B., Yau, S.-T., (eds.) *Mirror Symmetry II*, Studies in Advanced Mathematics. American Mathematical Society, 1997, Providence. [A collection of articles containing the latest results on mirror symmetry.]

III. Theory of Condensed Media

1. De Gennes, P.G., *The Physics of Liquid Crystals*, Clarendon Press, Oxford, 1974. [A clear and accessible non-specialized explication of the physics of liquid crystals.]

2. [VW] Vollhardt, D., Wölfle, P., *The Superfluid Phases of* ^{3}He. Taylor and Francis, London, 1988. [A detailed analysis of ^{3}He.]

3. [K1] Kléman, M., *Points, Lines, and Walls in Liquid Crystals, Magnetic Systems, and Various Ordered Media*, John Wiley & Sons, Chichester, 1983.

4. [K2] Kléman, M., "Defects in liquid crystals," *Rep. Prog. Phys.*, **52** (1989), 555–654. [The book [K1] and this review give a classification of defects in different media. All the basic methods are presented, including topological methods.]

5. [CP] Chakraborty, T., Pietilainen, P., *The Quantum Hall Effect, Fractional and Integral*, 2nd ed., Springer-Verlag, Berlin, 1995. [A survey of the current state of the problem accessible to nonspecialists.]

6. [LS] Lipowsky, R., Sackmann, E., (eds.) *Structure and Dynamics of Membranes*. Elsevier, Amsterdam, 1994. [A collection of papers on the theory of membranes. All the main areas of research are presented.]

7. [LPS] Le, T.Q.T., Piunichin, S., Sadov, V., "Geometry of quasicrystals," *Russ. Math. Surveys*, **48** (1993), 41–102. [This paper studies the problem of classifying quasicrystals.]

8. [SO] Steinhardt, P., Ostlund, S., (eds.) *Physics of Quasicrystals*. World Scientific Publishing Company, Singapore, 1987. [A collection of fundamental papers on the physics of quasicrystals.]

9. [JR] Joets, A., Ribotta, R., "Caustics and symmetries in optical imaging. The example of connective flow visualization," *J. Phys.* (France), **I** (1994), No. 4, 1013–1026.

10. [JMR] Joets, A., Monastyrsky, M., Ribotta, R., "Ensembles of singularities generated by surfaces with polyhedral symmetry," *Phys. Rev. Lett.*, **81** (1998), No. 8, 1547–1550. [A study of caustics in liquid crystals.]

11. [NM] Novikov, S.P., Mal'tsev A.Ya., "Topological phenomena in normal metals," *Russian Physics—Uspechi*, **41** (1998), No. 3, 231–240. [The newest invariants of Fermi-surfaces are found.]

12. [PP] Patashinskii, A., Pokrovsky V., *Fluctuation Theory of Phase Transitions*, Pergamon Press 1979.

IV. Theory of Gauge Fields

1. [W] Weinberg, S., *The Quantum Theory of Fields*, Vol. 1 (1995); Vol. 2 (1996) Cambridge University Press, Cambridge, U.K. [A basic course in field theory, including an analysis of topological particles.]

2. [Po] Polyakov, A.M., *Gauge Fields and Strings*, Harwood, London, 1987. [An informal exposition of the basic ideas and methods of modern field theory.]

3. [O] Okun, L.B. *Particle Physics: The Quest for the Substance of Substance*. Harwood, London, 1986. [A survey of the basic concepts of elementary particle physics, including the theories of strong and weak interactions based on quarks and extensive experimental material, aimed at nonspecialists.]

4. [GSW] Green, M.B., Schwarz, J.H., Witten, E., *Superstring Theory*, Vols. 1 and 2. Cambridge University Press, Cambridge, U.K., 1987. [A systematic exposition of the basic concepts and methods of string theory.]

V. Nonlinear Equations

1. [ZMNP] Zakharov, V.E., Manakov, S.V., Novikov, S.P., and Pitaevskii, L.P., *Theory of Solitons*, Plenum, New York, 1984. [A systematic exposition of the theory of solitons. It presents the method of inverse scattering, which is the main tool for investigating nonlinear integrable equations.]

2. [N] Newell, A.C., *Solitons in Mathematics and Physics*. [A beautiful introduction to the subject. It contains many interesting examples of applications of the theory of solitons and the history of the origins of the concept of a soliton.]

3. [Sp] Springer, G., *Introduction to Riemann Surfaces*, Addison–Wesley, Reading, MA., 1957; 2nd ed., Chelsea, New York, 1981.

4. Filippov, A., *The Versatile Soliton*, Birkhäuser Boston, Cambridge, MA, 1999.

VI. Knots and Links

1. [At] Atiyah, M., *The Geometry and Physics of Knots*, Cambridge University Press, Cambridge, U.K., 1990.

2. [Bi] Birman, J., "New Points of View in Knot Theory," *Bull. Amer. Math. Soc.*, **28** (1993), No. 2, 253–287.

3. [Jo] Jones, V.F.R., *Subfactors and Knots* (CBMS No. 80), American Mathematical Society, Providence, 1991.

4. [Ka] Kauffman, L., *Knots and Physics*, 2nd ed., World Scientific Publishing, Singapore, 1994.

5. [Ko] Kontsevich, M., "Vassiliev's Knot Invariants," in: *Advances in Soviet Mathematics*, **16**, Pt. 2 (1993), 137–150.

6. [MR] Moffatt, H.K., Ricca, R.L., "Helicity and Călugareanu Invariants," in: *Knots and Applications*, L.H. Kauffman, ed., World Scientific Publishing, Singapore, 1995.

7. [MS] Monastyrsky, M., Sasorov, P., "Topological invariants in magnetohydrodynamics," *JETP*, **93**, No. 10 (1987).

8. [N] Nechaev, S.K., *Statistics of Knots and Entangled Random Walks*, World Scientific, Singapore, 1996.

9. [Ro] Rolfsen, D., *Knots and Links*, Publish or Perish, Berkeley, 1976.

10. [Tu] Turaev, V., *Quantum Invariants of Knots and 3-Manifolds*, Walter de Gruyter, Berlin, 1994.

11. Murasugi, K., *Knot Theory and Its Applications*, Birkhäuser Boston, Cambridge, MA, 1996.

VII. Popular and Biographical Books

1. [H] Hawking, S.W. *A Brief History of Time from the Big Bang to Black Holes*. Bantam, Toronto, 1988.

2. [GB] De Gennes, P.G., Badoz, J. *Fragile Objects: Soft Matter, Hard Science, and the Thrill of Discovery*, Springer-Verlag, New York, 1996.

3. [LS] Lederman, L.M., Schramm, D.N. *From Quarks to Cosmos*. 1995.

4. [U] Ulam, S. *Adventures of a Mathematician*, Charles Scribner's Sons, New York, 1976.

Index